# A DECISION FRAMEWORK FOR WETLAND-RIVER BASIN MANAGEMENT IN A TROPICAL AND DATA SCARCE ENVIRONMENT

# A Decision Framework for Integrated Wetland-River Basin Management in a Tropical and Data Scarce Environment

DISSERTATION
Submitted in fulfillment of the requirements of
the Board for Doctorates of Delft University of Technology
and of the Academic Board of the UNESCO-IHE Institute for Water Education
for the Degree of DOCTOR
to be defended in public on
Monday 17 December, 2012 at 15:00 hours
in Delft, the Netherlands

*by*

**Mijail Eduardo, ARIAS HIDALGO**
born in Guayaquil, Ecuador

Bachelor of Civil Engineering,
Escuela Superior Politécnica del Litoral (ESPOL), Guayaquil, Ecuador
Master of Water Science and Engineering (with distinction),
UNESCO-IHE, Institute for Water Education, Delft, the Netherlands

This dissertation has been approved by the supervisors:
Prof. dr. ir. A. E. Mynett
Prof. dr. ir. A.B.K. van Griensven

Members of the Awarding Committee:

| | |
|---|---|
| Chairman | Rector Magnificus, Delft University of Technology |
| Vice-chairman | Rector, UNESCO-IHE |
| Prof. dr. ir. A.E. Mynett | UNESCO-IHE / Delft University of Technology, supervisor |
| Prof. dr. ir. A.B.K. van Griensven | UNESCO-IHE/ Vrije Universiteit Brussel, supervisor |
| Prof. dr. ir. U. Shamir | Technion Institute, Israel |
| Prof. dr. ir. P.L.M. Goethals | Universiteit Gent, Belgium |
| Prof. dr. ir. P. van der Zaag | UNESCO-IHE / Delft University of Technology |
| Prof. dr. ir. M.J.F. Stive | Delft University of Technology |
| Prof. dr. ir. G.S. Stelling | Delft University of Technology (reserve member) |

CRC Press/Balkema is an imprint of the Taylor & Francis Group, an informa business

Published by:
CRC Press/Balkema
PO Box 11320, 2301 EH Leiden, The Netherlands
e-mail: Pub.NL@taylorandfrancis.com
www.crcpress.com - www.taylorandfrancis.com

ISBN 978-1-138-00025-4 (Taylor & Francis Group)

*"Tell me and I forget. Teach me and I remember. Involve me and I learn!"*

**Benjamin Franklin**, US statesman, scientist,
philosopher, writer and inventor (1706-1790)

*To my mother, my father,
my wife and my daughter*

# Contents

# List of Figures

# List of Tables

# Acknowledgments

The road was not free of thorns, but has taught me a lot. And I could not end my time here in Delft without being grateful. Indeed, it is one of the most important things human beings should learn and never forget. First of all, I would like to thank Our Lord and His beloved Mother for having conducted me through this wonderful path. In their mysterious ways, they have shown to me that not always the most obvious outcome or the things we wish is what we really need. I infinitely thank them for having comforted me all these years, especially during hard times.

I would like to express my sincere gratitude to my supervisor, Prof. Arthur Mynett for his vision, wisdom and support in many ways. Since the time of my MSc. research when he chose me as one of his students, he has given me confidence and most importantly, he taught me that successful research is often based on "learning from of mistakes". In this regard, he made me appreciate even those moments when things did not work out the way I had hoped or seemed hopeless altogether. In addition to his guidance, I would like to thank him for his financial assistance, both during my MSc. research within his Strategic Research Department in Delft Hydraulics (with particular gratitude to both secretaries Frances Kelly and Jitka van Pommeren) and later during the final stages of my PhD research. Surely, without him, I would not have finished this important episode of my life.

My co-supervisor, Dr. Ann van Griensven, deserves a lot of acknowledgements as well. She approached me when I had finished my MSc thesis and offered me the opportunity to work within the WETWin project. Besides, she has been fully supportive during all the crazy scientific enterprises I have been involved in, and gave me good ideas and suggestions that undoubtedly enriched this work. At UNESCO-IHE, my particular thanks go to Dr. Biswa Bhattacharya and Dr. Giuliano Di Baldassarre who unconditionally shared their knowledge and patience, guiding me through the roads of hydrologic and hydrodynamic modeling.

I wish to deeply thank the WETWin project for having funded this thesis during its first three years. In a special way I am grateful with Dr. István Zsuffa, Dr. Jan Cools and Prof. Uri Shamir for providing me the inception on the project goals and their valuable suggestions, comments throughout the fruitful progress meetings in three continents! Institutionally, I thank the ESPOL-CADS staff for their logistic support during the data collection campaigns in Ecuador, particularly to Dr. Pilar Cornejo and Dr. David Matamoros. Despite the difficulties during the data collection process, I discovered a talented group of people from whom I learned a lot and thus let me grow as a professional. Enormous thanks go to Patrick Debels who was my mentor during the first year of my research. Gonzalo Villa-Cox also deserves my infinite gratitude for sharing his deep insight in econometrics and statistics as well as for being a good friend and a skillful project colleague. In the same way I deeply acknowledge the friendship and ever present support that

Gabriela Alvarez, Juan Carlos Pindo and Fernando Jarrín provided in the areas of water quality modeling, data mining and hydrological modeling, respectively. Our fantastic team was completed with the ESPOL-CIEC staff, especially Dr. Gustavo Solórzano and Ramón Villa-Cox. They provided fruitful suggestions to my work as well as were being tough opponents on the chess board!

UNESCO-IHE and Delft have been, undoubtedly, like a dreamful experience for me. During six years since I arrived for my MSc (they passed like the blink of an eye!), I have found so many good people and most importantly, friends; truly, a second family. Either from Latin America or the rest of the world, my sincere acknowledgements go to those who, in one way or another, have strengthen me and shared special moments. I hope not to forget any of them: Gabriela, Belén, Verónica, Jairo, María, Sofía, Anh, Evelyn, Christian, Gerald, Hong Li, Winnie, Carlos, Alejandro, Helga, Benly and many others (I apologize in advance for not mentioning all of you!). Special thanks to Davide, Klaas, Roberto, Emeline, Fredy, Alonso, Aklilu, Stefan, Omar, Maurizio, Diego, Neiler and Jonathan for those philosophic discussions during the so famous *third halves* after the freezing football training sessions. I am also grateful to with my fellow PhD colleagues: Raquel, Patricia, Girma, Linh, Maribel, Adrian, Loreen, Yuquing, Xuan, Assiyeh, Micah and Kun for their sincere friendship and useful suggestions. In a similar way, I thank to the rest of staff members at IHE, particularly to Wendy Sturrock, Jolanda Boots, Maria Sorrentino, Jos Bult, Ineke Melis and Guy Beaujot. Finally, my friends in Ecuador deserve a lot of appreciation too. This work is also dedicated to them: Darío, Leonardo, Silvana, Nacho, Eduardo, Carlos, Juan, Tere, Betsy, Fanny, Rocío, Johanna. Mis amados amigos, ¡nunca los he olvidado!

In a special way I show my everlasting gratitude to my family in Guayaquil, Ecuador. I remember now my dear sisters, Karina and Rosita, my brother Luis Antonio and my parents-in-law, Irina and Jaime, and sister-in-law Carolina who gave me heartfelt support during this stage. Similarly, I deeply acknowledge my gratitude to my aunts, uncles, cousins and grandmothers, for all their love and blessings. I deeply express my admiration and gratitude to my parents, Rosa and Luis, with whom I have an eternal debt, for all the lessons of life that granted to me. Apart of their endless love, they taught me that with hard work, honesty and devotion, everything is possible. Some dear people with whom I started this adventure are no longer physically amongst us, but remain always in my mind and heart. With special affection I would like to remember my father Luis, my grandma Pura Delia, my brother-in-law Angel, my uncles Ignacio and Alex and my friend P. Guillermo. I miss them, but hélas! C'est la vie and we must go on.

Last but not least, my greatest feelings go to my dear wife, Nina. She has been an endless source of love, inspiration, patience and support, especially when I was far away from home. Along with our beloved daughter, Valeria Alexandra, both make every day a living dream. Мои любви, я вас люблю всем сердцем: сильно, сильно, крепко, крепко! Я вернусь и сейчас мы будем вместе и навсегда!

Delft, the Netherlands, 17 December 2012

# Summary

Wetlands are probably one of the most astonishing ecosystems across the Earth. They are present in almost every latitude zone, from the arctic tundra's to the tropical wetlands near the equator. Being important parts of watershed ecosystems, wetlands often play a key role in fauna-flora conservation, habitat preservation and attracting recreation. But they also provide important ecosystem services related to water buffering, flood flow regulation water pollution control and water quality improvement.

Although traditionally wetlands were often considered separately from riverine systems, nowadays an integrated approach is becoming common practice in wetland-riverine watershed analysis and management. Such overall environmental approach implies not only an adequate representation of relevant bio-physical parameters, but also of socio-political and economic indicators. This is where computer-based modeling and decision support becomes extremely helpful. Currently, there are several complex simulation tools available that try to represent the different processes taking place in river catchments including wetland systems. However, more complexity may also lead to more uncertainty and can become *too much* for a particular need, in particular in multi-criteria evaluation. Therefore, a balance has to be established between simplicity and sophistication before building any model for decision support.

Based on these considerations, the primary goal of this research is to elaborate a simple but useful integrated framework or methodology for a coupled wetland-catchment environment, containing (i) both quantitative and qualitative approaches; (ii) incorporating stakeholders' feedback; (iii) taking into consideration the pressures on the system; and (iv) evaluating relevant management solutions for good decision making. From this main goal, the following specific objectives are derived:

1. to organize the available information within the considered wetland catchment and adjacent river(s) and collect additional data if possible; this process should take in consideration both qualitative and quantitative aspects of the wetland-river system to be studied;
2. to perform a pattern characterization and gap analysis (when required) of river discharge time series, as a pilot step prior to a rainfall-runoff simulation;
3. to establish a set of indicators and define which will be assessed in a quantitative or a qualitative way, depending on the data availability. New sources of rainfall data are to be explored as an alternative for conventional hydrological modeling. Ultimately, a modeling framework is to be developed to quantitatively explore the effects of proposed management solutions;
4. to invoke expert elicitation when evaluating proposed management solutions that require additional information or qualitative appreciation;

5. to achieve a final ranking of management solutions via a decision support process where the preferences of scientists, stakeholders and decision makers are all taken into account.

The EU-funded WETWin project is one of the recent efforts to bridge the gap between theory and practice. Its main goal was to stimulate a positive reform on the role of wetlands in the framework of integrated water resources management. To this end, WETWin considered several aspects to be of paramount importance, notably (i) actively involving stakeholders, (ii) enhancing institutional capacities to adopt and implement management options, (iii) achieving a balanced tradeoff between ecological services and human interests, and (iv) exchanging South-to-South knowledge and experience obtained from different WETWin case studies.

The Abras de Mantequilla wetland in Ecuador (a Ramsar site) was selected as the case study to implement the proposed methodology. This system is located in the middle of the Guayas River Basin (34000 $Km^2$), one of the main hydrographic ecosystems in Ecuador. Across the basin, three major rivers are worth mentioning: the Daule, the Vinces and the Babahoyo. Despite the overall low degree of human intervention, a DPSIR analysis (Drivers, Pressures, States, Impacts and Response) identified two main pressure drivers on the system (i) the major infrastructure works at the basin scale, planned by SENAGUA (Water Ministry of Ecuador), namely the Baba multipurpose dam construction and the DauVin irrigation projects, and (ii) the land use degradation at the wetland scale. In order to assess the possible effects of these events, a set of management solutions at the local scale have been explored and compared with the Business As Usual scenario. The BAU scenario included the probable climatic variations as well as the modifications that the Baba and DauVin projects may exert on the system. The management solutions mainly involved water retention measures in the wetland during the dry months, and different degrees of land use changes.

The recurrent lack of data was a major obstacle in identifying the system characteristics, building appropriate models, and ultimately the decision making process. Streamflow characterization is an important aspect of river basin development, but remains a challenge, especially in developing countries. In this thesis a methodology has been developed to identify trends and to estimate the main characteristics of data gaps for river flow time series at stations sparsely scattered across the study area. The proposed technique makes use of a band pass filter according to Hodrick-Prescott, to transform the measurements into a Fourier series to estimate gap patterns in river discharge time series. Inside the filter process, two approaches were explored (i) block homoskedastic and (ii) heteroskedastic. The methodology was applied to river stations across the study area that, in general, had a daily resolution. The approach distinguished between the two seasons throughout a year. This relatively simple approach proved quite relevant when insufficient data (typical in developing countries) makes it difficult to apply hydrological rainfall-runoff models or even linear regression models based on nearby gauging stations. The results from the computational tool developed to compare patterns from estimated and original time series, provided quite adequate to resolve gaps of various lengths. It should be noted of course that this method never sought to match the real time series on a one-to-one basis, but mainly to reproduce the same pattern characteristics (mean trend, dynamic variability) and establish the bands for possible approximations and the observed values.

A modeling framework has been developed for the analysis of scenarios and management options in the Vinces catchment and the Abras de Mantequilla wetland. Basically, it has two main functions (i) to serve as a tool to characterize the system and (ii) to provide data in the wetland areas where information is scarce. In addition, the modeling framework was used to explore effects of potential measures to the identified DPSIR chains for the present case study. Two hydrological indicators were assessed, namely water quantity and water quality. The chain of simulation models included two rainfall-runoff models, one river routing and one water allocation model.

Results from the rainfall-runoff model in the Vinces catchment performed quite well at several measurement locations, despite the strong assumptions made. As a consequence, given the good comparison between modeled and measured runoff data in case of reliable rainfall input, some measured discharge observations in fact became *suspicious*. As an alternative, new technologies are being used to expand the coverage of conventional meteorological datasets. An example of these is the TRMM (3B42) satellite data. As long as one considers the bias, the type of rainfall and its limitation such as the spatial resolution (not currently applicable to small-scale studies), TRMM data can prove helpful to fill in data gaps. The spatial distribution of the annual rainfall data from TRMM to some extent showed some similarity to the pattern from the ground raingauges. Bias correction factors were calculated and, adopting a simple procedure, were spatially distributed, and thus used to improve the satellite TRMM data. Using an empirical, yet effective disaggregation method, it was possible to generate synthetic daily rainfall time series at the satellite spots. These artificial series were incorporated in the existent rainfall-runoff model to complement the ground-based input data and then to assess its performance. The results were quite comparable with those using only gauge information. Consequently, TRMM data can be resourceful in areas where there are no rain gauges such as the Andean foothills in the case of the Guayas River Basin.

Furthermore, a hydrodynamic simulation model was built encompassing the Vinces and Nuevo rivers as well as the wetlands. A river routing model used the flows from the Vinces' tributaries as part of its boundary conditions and then computed flows and water stages in the downstream area. The flows between the Nuevo River and the Abras were seasonal being very strong during the rainy season but in May turned to a stagnant point once the precipitation regime ended.
Finally, a water allocation model computed the distribution of volumes across the main rivers of the Guayas watershed. The expected increased climatic variations, the major hydraulic works as well as the proposed management options were assessed using this tool. Firstly, the model showed that the flows along the rivers might tend to increase noticeably after 2020, as a consequence of the incremental

trend in precipitation expected for the positive influence the Ecuadorian Coastal Region Secondly, it was confirmed that the Lulu and San Pablo rivers have a positive effect on the system, especially once the Baba Dam will start operating and even more so when climate changes may occur. Five management solutions (MS) have been developed and compared with the *Business as Usual* alternative. The simulations converged to the conclusion that alternative MS4 (water retention + better agricultural practices + moderate crop substitution + ecological corridors) was the most convenient to the system, given its high scores on water quantity and water quality impacts.

A fully integrated analysis could not be based exclusively on quantitative model simulations. In some cases insufficient data availability required additional input via expert judgment or *human feedback*. Also, for socio-economic indicators the opinion of stakeholders becomes crucial to complement the model-based hydrological perspectives that the management solutions initially were based on. Here, the Lickert scale served as the *Rosetta stone* to convert subjective perceptions into numerical values. For each indicator, a value function was constructed, either from other sources such as the WET-Health methodology, or by using particular mathematical expressions. A panel was composed containing different areas of expertise, to score the performance of each proposed management solution. The scores tended to be higher as the complexity of the MS increased, similar to the results obtained from the quantitative modeling approach.

The last stage of the proposed methodology was to incorporate the stakeholders' preferences and integrate all sources of information into a system to support further management decisions. The different outcomes from the analyses of the proposed management solutions were harmonized into an evaluation matrix by means of selected appropriate value functions. A set of weights, one for each indicator, representing the influence of each of the final ranking of management solutions, was derived from workshops with the stakeholders involved: local inhabitants and government officials. From the resulting ranking of management solutions, it was observed that the most elaborated alternative (MS5) was preferred amongst the authorities. On the other hand, the local stakeholders were of the opinion that small-scale crop substitution and some reforestation would be the best way to restore the wetland. However, some time in future they could well see more radical changes in land use cover, which could imply that MS5 may well become their choice later on. In this way, a balance could be reached between the most important ecological services the wetland provides and the various goals of the stakeholders involved. Ultimately, by allowing space for future negotiations among the actors, this methodology is amenable to continuous enhancement towards a better wetland and river basin management.

# Samenvatting

Wetlands behoren tot de meest verbazingwekkende ecosystemen ter wereld. Ze zijn alom aanwezig, van de tundra's bij de poolgebieden tot de tropische wetlands bij de evenaar. Als belangrijk onderdeel van ecosystemen in stroomgebieden spelen wetlands vaak een cruciale rol in het behouden van flora en fauna, het verschaffen van leefgebieden, en het voorzien in recreatiemogelijkheden. Maar zij voorzien ook in belangrijke ecosysteemfuncties met betrekking tot het opslaan van water, het reguleren van overstromingen, het zuiveren van afvalwater, en het verbeteren van de waterkwaliteit.

Hoewel wetlands in het verleden vaak werden beschouwd als afzonderlijke water-gebieden los van het riviersysteem, worden ze tegenwoordig gezien als integraal onderdeel bij het integraal beheer van stroomgebieden. Dit betekent dat niet alleen de relevante bi-fysische parameters worden meegenomen, maar ook de socio-economische indicatoren. Belsissingsondersteunende software systemen kunnen hierbij zeer behulpzaam zijn. Er zijn dan ook verschillende complexe systemen in omloop die evenwel om grote hoeveelheden gegevens vragen. En deze zijn niet altijd beschikbaar, zeker niet in ontwikkelingslanden. Vandaar dat het belangrijk is een balans te vinden tussen eenvoud en detail bij het opzetten van dergelijke modellen.

Op basis van deze overwegingen is het belangrijkste doel van dit onderzoek dan ook om een eenvoudig doch bruikbaar raamwerk te ontwikkelen dat integrale stroomgebieden (inclusief wetlands) kan representeren, waaronder (i) zowel kwantitatieve als kwalitatieve benaderingen; (ii) commentaren van betrokkenen; (iii) verschillende belastingen op het systeem; en (iv) relevante opties voor het nemen van goede beslissingen. Hieruit volgen de volgende specifieke doelstellingen:

1. de beschikbare informatie over wetlands en stroomgebieden samenvoegen en rangschikken en zo mogelijk van aanvullende informatie voorzien;
2. de karakteristieke patronen te herkennen en leemtes in gegevens in te vullen;
3. een aantal indicatoren te benoemen en aan te geven welke zullen worden gebruikt om alternatieven te beoordelen;
4. de kennis van experts te gebruiken om voorstellen te evalueren en zonodig aan te vullen;
5. uiteindelijk tot een rangschikking te komen van beheersalternatieven die recht doet aan de verschillende voorkeuren;

Het door de EU gefinancierde WETWin project is een van de recente ontwikkelingen om te trachten de brug tussen theorie en praktijk te slechten. Het belangrijkste doel was om de rol van wetlands bij integraal waterbeheer te (he)rkennen. Het WETWin project besloot de nadruk te leggen op verschillende aspecten die van cruciaal belang werden beschoud, nl. (i) het actief betrekken van belanghebbenden; (ii) het versterken van instanties die beslissingen moeten implementeren; (iii) het bereiken van een gebalanceerde  afweging tussen ecologische en menselijke belangen, en (iv) het uitwisselen van kennis en ervaring opgedaan in het WETWin project tussen de zuidelijke landen onderling.

Het Abras de Mantequilla wetland in Ecuador (een RAMSAR locatie) werd gekozen tot onderzoeksgebied om de voorgestane methodiek uit te proberen. Dit wetland is gelegen midden in het Guayas River Basin ($34000$ Km$^2$), een van de belangrijkste hydrografische ecosystemen in Ecuador.  Het gebied omvat drie rivieren: de Daule, de Vinces en de Babahoyo.

Hoewel redelijk onontgonnen, wees een DPSIR analyse uit dat het systeem te maken had met twee mogelijke bedreigingen: (i) de belangrijke infrastructurele werken die in voorbereiding waren bij SENAGA (het ministerie van water in Ecuador), namelijk de constructie van de Baba dam en de DauVin irigatieprojecten, en (ii) de achteruitgang van de grond samenstelling binnen het wetland. Om de mogelijke gevolgen na te gaan is een aantal beheersvarianten nagegaan en vergeleken met de huidige benadering. Aangezien de huidige benadering al rekening hield met zowel mogelijke effecten van klimaatverandering als met de op handen zijnde constructiewerkzaamheden, richtte het WETWin project zich vooral op het vasthouden van water in het wetland gedurende het droge seizoen, alsmede met het veranderen van het grondgebruik.

Een steeds terugkerend obstakel om het systeem goed te kunnen karakteriseren en modelleren was het gebrek aan gegevens. Afvoermetingen van rivieren zijn belangrijk om modellen te ijken, maar ontbreken vaak in ontwikkelingslanden, zo ook hier. Vandaar dat in dit proefschrift een methodiek is ontwikkeld om trends the kunnen analyseren en leemtes in beschikbare gegevens te kunnen invullen. De techniek is gebaseerd op een band-filter algoritme van Hodrick-Prescott waarmee metingen in Fourier componenten worden gesplitst om gaten in afvoertijdreeksen te kunnen dichten "in statistisch correcte zin".   Er zijn twee methodieken onderzocht om te filteren: (i) blok-homoskedastisch en (ii) heteroskedastisch. De methodieken zijn toegepast op meetstations in het studiegebied met tijdstappen van een dag. Er werden twee seizoenen onderscheiden per jaar, een nat en een droog zeizoen. Deze betrekkelijk eenvoudige benadering bleek afdoende wanneer gebrek aan gegevens (typisch voor ontwikkelingslanden) het niet mogelijk maakt om hydrologische regenval-afvoermodellen te gebruiken of lineaire regressie-modellen toe te passen. Op deze manier konden lacunes in meetgegevens van verschillende duur worden aangevuld. Hierbij moet worden opgemerkts dat deze methode niet bedoeld is om echte waarnemingen een-op-een te reproduceren, alswel om de karakteristieken eigenschappen (verloop van de gemiddelde waarde en variabiliteit daar omheen) te reproduceren en de waarschijnlijkheidsgrenzen grenzen te schatten.

Op basis hiervan is een raamwerk opgezet waarmee scenario's en management opties kunnen worden geanalyseerd in het Vinces stroomgebied en het  Abras de Mantequilla wetland.  Dit raamwerk heeft twee hoofdfuncties: (i) het watersysteem te karakteriseren en (ii) gegevens aan te vullen waar informatie niet of beperkt beschikbaar is. Daarnaast is het model raamwerk gebruikt om de effecten van mogelijke maatregelen te onderzoeken in de DPSIR benadering die in dit testgeval is gebruikt. Er zijn twee indicatoren onderzocht: water kwantiteit en water kwaliteit. De onderliggende modelsystemen bevatten twee regenval-afvoermodellen, een riviermodel, en een waterdistributiemodel.

De resultaten in het Vinces stroomgebied kwamen goed overeen met waarnemingen van verschillende meetstations, ondanks alle benaderingen die werden toegepast. Het is interessant om op te merken dat de goede overeenkomst tussen berekende en gemeten afvoergegevens ertoe leidde dat bij sommige waarnemingen *vraagtekens* konden woden geplaatst. Om dit te onderzoeken werden TRMM (3B42) satellietwaarnemingen gebruikt die het bereik van standaard beschikbare meteo gegevens kunnen vergroten. Ondanks beperkingen in resolutie kan op die manier toch een redelijke schatting worden gemaakt. De ruimtelijke verdeling van jaarlijkse neerslag gegevens uit TRMM gegevens leverde een vergelijkbaar patroon op als die op basis van meetstations op de grond.

Op basis van een eenvoudige aanpak werden correctiefactoren bepaald en werden TRMM gegevens omgezet naar synthetische tijdreeksen van dagelijkse waarnemingen op specifieke locaties. Deze reeksen werden als invoer gebruikt voor het bestaande neerslag-afvoermodel en de resultaten geanalyseerd. De uitkomsten bleken zeer vergelijkbaar met die op basis van alleen meetstations. Dit betekent dat het gebruik van TRMM waarnemingen een uitkomst kan zijn in gebieden waar geen meetgegevens beschikbaar zijn, zoals in de uitlopers van het Andes gebergte voor het stroomgebied van de rivier de Guayas.

Daarnaast is een hydrodynamisch model gebouwd voor het stroomgebied van de Vinces en Nuevo rivier, inclusief de wetlands. Het stroommodel voor de rivieren leverde de randvoorwaarden voor het benedenstroomse gebied. De rivierafvoeren waren het grootst in het regenseizoen maar belven zeer beperkt gedurende het droge jaargetijde, vanaf de maand Mei. De uitvoer van dit model leverde de randvoorwaarden voor  een waterverdelingsmodel in het stroomgebied van de Guayas. Op die manier konden effecten van klimaatverandering en de aanleg van bovenstroomse kunstwerken worden onderzocht, alsmede nut en noodzaak van mogelijke beheersmaatregelen.

De resultaten lieten allereerst zien dat de rivierafvoeren waarschijnlijk zullen toenemen vanaf het jaar 2020 ten gevolge van verwachte toename van regenval langs de Ecuadoriaanse kust.  Ten tweede bleek inderdaad dat de rivieren Lulu en San Pablo een positieve bijdrage leverden, met name wanneer de Baba dam in gebruik zal worden genomen en des te sterker wanneer klimaatverandering zou optreden. Op basis hiervan zijn vijf beheersmogelijkheden onderzocht en vergeleken met de bestaande aanpak. De uitkomsten convergeerden naar de conclusie dat alternatief MS4 (water vasthouden + betere landbouw technieken gebruiken + beperkte verandering in gewassenteelt aanbrengen + ecologische zones aanleggen) de beste uitkomsten gaf met name op het gebied van water kwantiteit en kwaliteit. In sommige gevallen waren onvoldoende gegevens beschikbaar en werd een beroep gedaan op de aanwezige *menselijke* kennis en ervaring in het stroomgebied. Dat geldt in nog sterkere mate voor socio-economische indicatoren waar de mening van belanghebbenden bepalend is.

De schaal van Lickert werd hier gebruikt als "steen der wijzen" om subjectieve meningen om te zetten in numerieke waarden. Er werd een panel samengesteld die verschillende expertisegebieden bestreek en die hun mening gaven over de verschillende voorgestelde management scenario's. De voorkeur bleek te gaan naar de meer complexe modellen, vergelijkbaar met de kwantitatieve model-benadering.

Het uiteindelijke doel van de voorgestelde methodiek was om de kwalitatieve benadering te combineren met de kwantitatieve, teneinde alle informatiebronnen te combineren in een enkel beslissingsondersteunend systeem. Daartoe werden alle uitkomsten genormeerd en in een beslissingsmatrix samengevoegd. De waarden van de normalisatieparameters werden vastgesteld aan de hand van gebruikers-bijeenkomsten met lokale bewoners en vertegenwoordigers van de overheid. Het eindresultaat leerde dat het meest uitgebreide alternatief (MS5) de voorkeur kreeg van de overheidsvertegenwoordigers. De lokale bewoners vonden dat op de korte termijn kleine veranderingen in teelt van gewassen samen met enige herbebossing voldoende waren om het wetland te herstellen. Wel zagen ze dat op de langere termijn drastischer maatregelen gewenst konden zijn en MS5 wel degelijk ook hun keus kon worden. Op die manier zou een balans gevonden kunnen worden tussen ecologische en economische waarden van de wetlands. De hier ontwikkelde methodiek kan ook in de toekomst worden gebruikt om veranderingen in voorkeuren van betrokkenen mee te nemen en te komen tot aanpassing van gebruiksfuncties voor een beter beheer van het stroomgebied, inclusief wetland.

# Chapter One

# I   Introduction

## I.1   Wetlands

Wetlands are undoubtedly one of the most astonishing ecosystems across the Earth.  They are present in almost every latitude zone, from the tundra to the equator in the tropic zone.  Wetlands encompass several kinds of surfaces including the typical swamps, marshes, fens, bogs and others.  Often perceived as dangerous sources of diseases like malaria, considerable wetland areas have been destroyed throughout history and with evident acceleration during the XX[th] century.  This happened to both inland (freshwater) wetlands (Fig. I.1-1 and Fig. I.1-2) as well as to coastal or salty marshes, such as the tropical mangrove swamps (mainly brackish water).

Wetlands have features that belong to both fully aquatic and entirely terrestrial ecological communities (Mitsch et al., 1988).   They are transition areas or ecotones which can play a role as permanent or temporary holders or exporters of organic – inorganic nutrients (Fig. I.1-3).  As a consequence of this, wetlands can sustain a broad range of species which often migrate between permanently flooded and dry places, thus becoming ecosystems with a very high biodiversity (Mitsch and Gosselink, 1986).  Hence, the study of wetlands is usually a little bit more complex than for other ecosystems, but challenging at the same time.  It necessarily implies a multi-disciplinary approach to simulate diverse processes (and thus perhaps many models) or evaluating several management options through a decision support system.  In recent times, Environmental Hydroinformatics (Mynett, 2002, 2004) has emerged as an important computer-based research field where wetland dynamics can be better simulated and analyzed using various data-mining techniques and numerical modelling approaches.

Fig. I.1-1 Inner Niger River delta, nearby Mopti, Mali.

Fig. I.1-2 Spreewald forest, Spree River, nearby Berlin, Germany.

Wetlands provide many services to the environment such as being a source of food for some civilizations. Examples are rice and corn crops in South America, or hay production in Northern Europe. In addition, wetlands are often receptors of wastewaters from neighbouring towns, villages and even cities (such as at Kampala, Uganda) and are regarded by some authors as the *"kidneys"* of the planet (Mitsch and Gosselink, 1986). Being important parts of a watershed ecosystem, wetlands play a key role in lowland catchments either as a water resource (surface and groundwater storage), flood flow regulation, recreation or providing fauna-flora conservation habitats (Kent, 2001), thus maintaining or increasing thus global biodiversity (Zacharias et al., 2005) and controlling water pollution (Hattermann et al., 2006) among others. In coastal and estuarine areas, storm surge effects can be significantly reduced by mangrove swamps. This emphasizes the potential of wetlands for flood control (against waves, hurricanes and high river water levels), thus becoming protectors of coastlines and catchments around the world (Bahuguna et al., 2008).

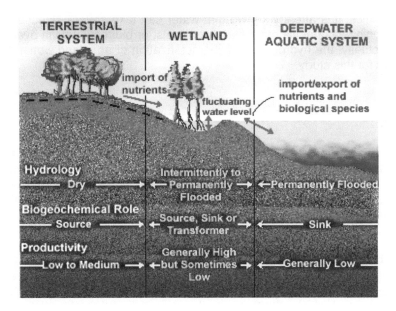

Fig. I.1-3 Wetlands sharing terrestrial and aquatic features. (Mitsch and Gosselink, 1986), http://ces.iisc.ernet.in/energy/Lake2002abs/ses1724.gif

A common conception is to consider wetlands and floodplains as quasi-synonyms (Blackwell et al., 2006). This assertion is based on 2 observations:

1. Wetlands usually constitute a large portion of a river floodplain; and

2. As mentioned above, wetlands often carry out multiple functions along a fluvial system, for instance  (i) flood reduction (Bullock and Acreman, 2003); (ii) buffer zone retaining harmful minerals and substances away from the river stream; (iii) capturing sediments and nutrients loads leading to ideal habitat for species; (iv) groundwater recharge zones and transition areas for surface and base flow exchanges. Although this definition can be applicable more generally, it is not yet commonly applied to highlands (Blackwell et al., 2006).

Several kinds of wetlands are found worldwide.  Among the most important classifications there are:  marine, estuarine, riverine, lacustrine (lake related) and palustrine (mostly inland swamps) (Kent, 2001).  This classification was developed in the 1970's by Cowardin for wetlands across United States (Cowardin et al., 1979).  Considering three relevant abiotic features such as (i) geomorphic settings; (ii) water resource, and (iii) hydrodynamics, wetlands can be classified based on geomorphology as well (Brinson, 1993).  Regarding geomorphic properties it is important to understand the wetland topography, quite important to reduce or accelerate runoff and thus determining the extension of a wetland (Mitsch et al., 1988), and its geology (mainly related to conductivity).  Finally, considering hydrodynamics, it is already known that water can take several directions inside a wetland:  unidirectional via creeks and channels, or bidirectional via overland flows mainly.  Actually, the hydrodynamic characterization of an inland wetland is one of the drivers for this study.

## I.2    Current status of wetland research

Many studies in the past and some even nowadays have traditionally addressed rivers and wetlands separately.  In fact, before 1996 the Ramsar Convention had not yet formally recognized a link between wetlands and river basins (Ramsar Secretariat, 2010).  Besides, in its initial stages the European Water Framework Directive (WFD) (EC, 2000a) itself was not clear about the domain of wetlands and their management policies (EC, 2000b).

Despite this early gap, recently there have been several successful attempts to jointly evaluate wetlands and rivers, at least from the modeling perspective. Illustrative examples are the ones conducted along the middle Seine River in France (Bendjoudi et al., 2002),  as well as the elaboration of guidelines for flood risk reduction across the European Union (Blackwell et al., 2006) and integrated models for riparian wetlands to emphasize their importance as *buffer zones* for the catchment (Hattermann et al., 2006).  In this way, a combined analysis of river-wetland systems is more and more required, especially facing the potential effects that climate changes may exert on water resources (Mynett, 2008).  This overall view implies the inclusion of social, eco-hydrological, biological and economical aspects and indicators (Chaves and Alipaz, 2007; Goosen et al., 2007).

And it is at the decision making stage that a deeper analysis is needed considering the interactions between rivers and wetlands. Many publications have widely covered the large-scale application of decision support tools for river basin management (Barrow, 1998; La Jeunesse et al., 2003; van Ast, 2000; Welp, 2001; Williams, 2001), giving marginal importance to the effects river basins can cause on wetlands. On the other hand, riparian wetlands have also been extensively studied but principally from a small-scale perspective and barely linking them to a river catchment (Cabrera, 2008; Goosen et al., 2007; Janssen et al., 2005; Kirk et al., 2004; Walters and Shrubsole, 2003). Due to this *restricted* view, the influence wetlands may exert on a larger riverine environment has been largely overlooked, most likely hampering proper decision making. Consequently, this mismatch between river and wetland analyses might have led scientists, stakeholders and particularly decision makers towards incomplete or poor management guidelines and policies. This was one of the drivers to conceive the WETWin project, as described hereafter.

## I.3   The WETWin project

### I.3.1 Concepts and scope

One of the most important and recent endeavours focused on the interaction between wetland-river ecosystems has been the WETWin project, 7th Framework Program – EU-FP7 (Zsuffa, 2008). Several practical problems and drivers world-wide motivated this initiative. To list a few:

1.  Notwithstanding the Ramsar Convention efforts to protect and regulate the management policies in wetlands, many of these ecosystems in developing countries are subject to unsustainable management practices or weak management policies from the local authorities (Zsuffa, 2008).

2.  The separate ways in which scientists, decision makers and stakeholders engage their activities towards a common water management goal.

3.  The lack of stakeholder involvement being a key issue to achieve sustainable benefits for the wetland communities.

4.  Wetlands, in the context of a river basin, are of paramount importance. This relevance is due to their manifold roles as agents for water storage, nutrient retention, ecological functions for the protection of the biosphere, etc.

5.  Adaptive management procedures and mitigation measures are highly required for integrated wetland-river catchment systems. The Millennium Development Goals (MDGs) are nowadays more and more difficult to

achieve for a growing population given the incremental pressure on water use, waste-water effluents, climate changes, etc.

6.  A multi-criteria approach is fundamental to attain a tradeoff amongst the wide-ranging distinct wetland functions and thus to deal with the different interests of the involved stakeholders.

7.  A stronger cooperation between southern countries is a challenging opportunity to share common experiences, assess differences and expand the current limitations.  This South-South twinning may constitute an innovative alternative to the traditional north-south interaction, and is one of the focal points of the WETWin project.

In general, the WETWin initiative aimed to stimulate a positive reform on the role of wetlands in the framework of integrated water resources management, taking into consideration not only quantitative factors, but also other aspects of paramount importance such as the social component or the work under data scarcity conditions.  To achieve this integrated way, specific targets were devised closely related to each of the seven considerations mentioned above (Zsuffa, 2008).

## I.3.2 Case studies and work packages in the WETWin project

A total of seven study areas have been selected for the project (Fig. I.3-1). All of them are inland wetlands and are related in different ways to a river basin. They also have differences which may allow a twinning stage at the end of the project. Some locations are distinguished by their size.  For instance, the inner Niger River delta (Mali) is the largest wetland case with more than 30000 $Km^2$. Pollutant problems, derived from human (e.g. the city of Mopti), industrial and irrigation waste disposal, and high evaporation rates may be the most significant hazards for the site (Kone et al., 2002; Zwarts and Diallo, 2002).  On the other hand, the Gamampa wetland in South Africa has a surface of around 1 $Km^2$ where the main pressure might be the rapid expansion of agriculture schemes.  The Namatala and Nabajjuzi wetlands are not only associated within the White Nile River catchment but also more specifically with the Lake Victoria and the urban growth of Kampala, capital city of Uganda.  Hence, their major problems seem to be sewage discharges from nearby cities and soil erosion.  In addition major challenges might be the increasing of fish production in the lake and wetlands and the use of papyrus as nitrogen retention agents (Kaggwa et al., 2008; van Dam et al., 2007). The European wetlands in consideration (e.g. Gemenc, Lobau) are characteristically floodplains next to a river course (the Danube).  There, not only modeling exercises but also integrating experiences were developed to tackle situations such as over-drainage, sedimentation, habitat destruction and landuse degradation (Hein et al., 2004; Hein et al., 2005).  Similar is the case of the Spreewald forest in Eastern Germany (Elbe-Havel-Spree system) where there is a severe problem of water quality possibly due to mining activities, low flows, and waste-water discharges flowing backwards from the city of Berlin (Wattenbach, 2008).

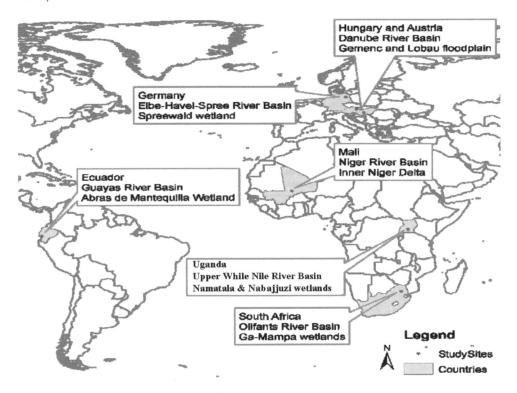

Fig. I.3-1 Case studies within the WETWin project.

Finally, the South-American case (the one considered in this thesis) is the Abras de Mantequilla wetland system (AdM) in the frame of the Guayas River Basin (GRB) in Ecuador. A set of drivers, pressures, states, impacts and responses for this environment were identified and discussed as motivation for the overall analysis. At a first glance, some potential drivers might be some major infrastructure works, planned by the national authorities, as well as local-scale landuse degradation, a common situation in the Ecuadorian lowlands. Yet another triggering conflict for this case is the low availability of data. Given a set of ecological services that the AdM may provide and the several concerns from the different stakeholders involved, a tradeoff analysis should be setup. To facilitate this, a group of management options, ranging from water storage to landuse substitution and reforestation, is to be explored. From that, a ranking of possible solutions may give an idea of the most *suitable* alternatives for this case, despite the default impossibility to fully satisfy all goals at once.

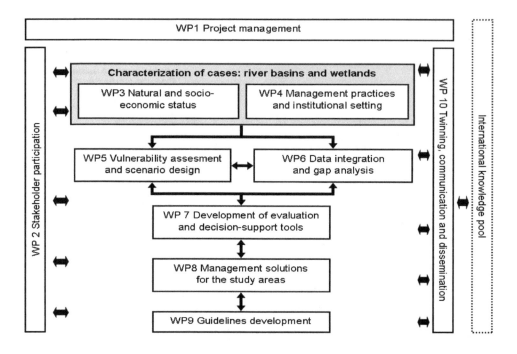

Fig. I.3-2 WETWin project flow and work package structure (Zsuffa, 2008).

In order to reach its objectives and to tackle the particular conditions of each case study, the project comprised ten work packages (WP) which attempted to cover all relevant fields from the water management perspective (Fig. I.3-2). The flow started with a management stage (WP1) and finalized with the elaboration of management guidelines (WP9), continuously fed back by stakeholder involvement (WP2) and supported by the dissemination and communication with the external scientific and public community (WP10). The other work packages, related with this thesis involved:

• WP3: Natural and socio-economic status. Its aim was to perform an overview on the availability of data and the socio-economic description of the current condition of the wetland area and its surroundings (nearby river catchments).

• WP4: Management practices and institutional settings. The main target was to evaluate the institutional capacity on integrated water resources management for each of the case studies.

• WP5: Vulnerability assessment and scenario design. Particularly, for the Ecuadorian case study, a composed scenario including the effect of climate changes and the major infrastructure works was foreseen.

- WP6: Data integration and gap analysis. A database was built up with the available data (WP3) and the data collected during the project. Another goal was to undertake a gap analysis when required.

- WP7: Development of evaluation and decision-support tools. Several modeling techniques and qualitative approaches, depending on the case study's pressures, needs, and environment were proposed, based on their applicability. These were the basis for the formation of a decision support system.

- WP8: Management solutions for the study areas. More than the definition which was part of early stages of the research, this entailed the final ranking process obtained through the decision support system.

## I.4   The balance between simplicity and complexity in water modeling

Since modeling is a key tool in the decision-making framework, it is important to consider how complicated a model should be. In general, models have traditionally been developed as a simplification of the reality. This means that the less simplified the model, the better or the more true it is. Nonetheless, this also implies more complexity since any *improved* simulation would have to deal with a higher number of variables / scales and thus more sources of uncertainty *(the complexity paradox)* and higher computational demands that ultimately may not entirely help (Gutjahr and Bras, 1993); in fact, it might be harder and harder to demonstrate that such model may lead to *the real world* (Oreskes, 2003). Modelers should first define the boundaries and limitations of their models before setting them up. This is not an easy question to solve since the ambition to mimic reality at all cost is always present. But when there is data scarcity this puzzle might become less difficult to solve, because the model complexity should be consistent with the availability of data and resources and the model targets (Clement, 2011; Mynett, 2002, 2004).

There is a distinction between studies that target scientific discovery or scientific breakthrough and those which aims to more practical applications. In the former case, models tend and need to be very sophisticated, for instance, when understanding at detail the interactions between soil and water in the unsaturated zone or when forecasting with high level of certitude an upcoming rainfall or a flood event. Such wetland models are helpful tools to describe, for example a strong interaction between wetlands and rivers (Krysanova et al., 1998) but also between overland flow, interflow and baseflow (Krause et al., 2007). For this kind of research, it has been discussed (Clement, 2011; Cunge, 2003) that very simple models may add more errors to the inherent uncertainty and thus they might become poor representations of what is really going on.

Secondly, there are the decision making or policy-modeling studies. In these situations scientists and engineers are commonly forced to use what is at hand or to choose between a very limited set of modeling options. Regardless of these inherent limitations, such models can still play a relevant role in supporting decision making. To achieve this role, model simulations can be very complex as long as they do not mislead to impractical, biased or quite complicated choices when contrasting their quantitative results with e.g. the social component. Noteworthy is that both *successful* and *futile* representations can provide precious lessons about the concepts and dynamics of the processes (Hunt and Welter, 2010). This learning process might be sometimes more useful than the outcomes of a complex model that cannot furthermore be applicable on the field, assuming of course that such sophisticated tool can be designed. Certainly, as it is shown in Fig. I.4-1, modeling does require time and money and it is justified as long as it does not cause the declining on the value of the investment for policy making (Carey and Zilberman, 2002; Clement, 2011).

Finally, the low capability to properly link various targets between stakeholders, scientists and decision makers was a major issue of research investment in earlier experiences within the European Union (Zsuffa, 2008). Hence, the necessity to balance what is invested in modeling referred to what is required and perhaps how much it might cost; all in favor of including other important data resources, such as expert elicitation.

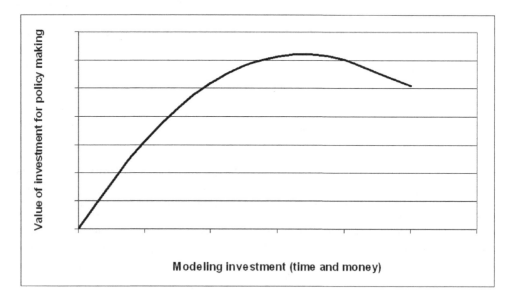

Fig. I.4-1 Modeling investment vs. the value of investment for policy making (Clement, 2011).

## I.5   Research questions

Based upon the aforementioned considerations, an overarching research hypothesis arises:

***Can an integrated framework of analysis for wetland-river systems be developed under data scarcity conditions, which will still facilitate the decision-making process on achieving sustainable wetland management?***

In addition, around this main question there are several issues that lead to complementary research questions, as follows:

Ecohydraulic issues:

1.  What is the interaction between the AdM wetland and its adjacent river (sub) catchment have on each other?  What is the potential of the wetland as water storage element?  What is the water quality level of the wetland?

2.  What can be suitable or alternative simple tools or approaches to deal with data scarcity and thus help in the model setup process?

3.  How applicable and informative are modeling tools for a river-wetland integrated analysis in data scarce areas?

Expert elicitation issues:

4.  If experts are consulted about possible management options, what will be their opinions?
5.  How different are the perceptions about wetland management amongst the different sorts of stakeholders?


Decision-making issues:

6.  What are the most probable scenarios and best management strategies for the wetland and river catchment in study in a near future?  How the wetland-river catchment system might behave should these events take place or in case the present conditions continue?

7.  How difficult can be to apply a decision support system under data scarcity conditions?

8.  How relevant is the wetland management in the river basin? How relevant are the river basin scenarios for the wetland management? How different are the management solutions for today compared to the ones in a nearby future?

## I.6  Research objectives

In order to solve these inquiries, the primary goal of this study would be *to elaborate a simple but useful integrated framework or methodology for a coupled wetland-catchment environment, covering both the quantitative and qualitative approaches under data scarcity conditions, incorporating stakeholders feedback, taking into consideration the pressures on the system and finally evaluating relevant management solutions for good decision making.*

The specific objectives are listed as follows:

1.  To organize the available information within the wetland catchment and the adjacent river(s) and collect new relevant when possible. This process takes in consideration the qualitative and quantitative aspects of the natural status, features and driver identification of the wetland and river catchment in study.

2.  To perform a pattern characterization and gap analysis (when required) of river discharge time series, as a pilot step prior to a river routing model.

3.  To establish a set of indicators and define which will be assessed in a quantitative or a qualitative way, depending on the data availability. As a complement, explore new sources of rainfall data for the hydrological modeling. Finally, characterize the system through a modeling framework and quantitatively explore the system behaviour facing the scenarios and the performance of the management solutions that are proposed.

4.  To assess the different opinions that the expert elicitation process can offer when evaluating the performance of each proposed management solution.

5.  To determine a final ranking of management solutions via a decision support system, where the feedback of scientists, stakeholders and decision makers is taken into account.

## I.7  Thesis outline

This research is outlined as follows:

Chapter I provides the introductory ideas, a general background, a brief description of the WETWin project, a discussion on complexity in modeling as well as defining the main objectives of the thesis and the research questions.

Chapter II describes in detail the case study on the Abras de Mantequilla wetland system, in the context of the Guayas River Basin in Ecuador, as a driver factor to develop the present study. It also describes the data collection process.

Chapter III shows a methodology to characterize patterns and estimate gaps along streamflow time series. For these targets, the characterization stage of the procedure is based on the Hodrick-Prescott filter (HPF) whereas the gap estimation uses Fourier series. The tool was applied to some river stations across the study area.

Chapter IV introduces a model cascade or framework. This quantitative analysis included the setup, calibration and results of two rainfall-runoff models, one river routing simulation and one water allocation model at a basin/wetland scale. This section also included a comparison between space borne and ground-based rainfall data and its posterior use as input data for the catchment hydrological model.

Chapter V goes through an expert elicitation process as a qualitative analysis tool to complement of the modeling framework of Chapter IV. The concepts of the Lickert scale and the value functions are incorporated following the opinion of the different consultants about the potential performance of the proposed management solutions.

Chapter VI covers the implementation of the decision support system (DSS) on the Abras de Mantequilla case study. This section made use of the analysis and evaluation matrices and the feedback from stakeholders and decision makers to come up with a final ranking for the management solutions. The chapter ends with a discussion on the performance of the *winner* choices.

Chapter VII presents the main conclusions and final recommendations.

Finally, the appendix A elaborates on an example of the questionnaire used to retrieve information from the consulted scholars and experts. Appendix B shows a list of the main socioeconomic and institutional indicators employed for the qualitative analysis.

$$\mathcal{C}hapter\ \mathcal{T}wo$$

# II   The Abras de Mantequilla case study

## II.1   The Guayas River Basin

The Guayas River Basin (GRB, 34000 $Km^2$) is located in the middle of the Ecuadorian coastal region (Fig. II.1-1). It is one of the most important areas in Ecuador, in terms of production. Three main activities take place within the basin, namely, urban-industrial development, agriculture and aquaculture (Falconi-Benitez, 2000; Southgate and Whitaker, 1994). More than 68% of the national crop production originates from this watershed (Borbor-Cordova et al., 2006; Cornejo, 2009).

Around 4.8 million inhabitants dwell in the watershed comprising part of 9 provinces, mostly in urban areas (Debels et al., 2009). Several cities and small towns are located along the river network being Guayaquil (3 million inhabitants), Quevedo, Babahoyo, Daule and Vinces (40000 inhabitants) amongst the largest and most populated. Ground elevations range across the basin from 0 up to 6310 meters above sea level (masl) at Mount Chimborazo. In this respect, four topographic sub-zones can be identified: the Guayas estuary (0-7m), the lowlands (7-50m), uplands (200-1500 masl) and highlands, on the Andean hill slopes, above 1500m (Borbor-Cordova et al., 2006).

Three major tributaries feed the Guayas River: the Daule, the Vinces and the Babahoyo Rivers. The Daule and the Babahoyo join waters north of Guayaquil City to form the Guayas, around 60 Km upstream of its mouth in the Gulf of Guayaquil (Pacific Ocean). The Guayas river has an annual discharge of about 30000 million $m^3$ (CEDEGE, 2000). The Daule River has a maximum average discharge of 1040 $m^3/s$ and flows southwards along the occidental flank of the basin. The Babahoyo River reaches up to 2100 $m^3/s$ on average in March. When an El Nino phenomenon occurs, the Guayas River may discharge more than 5000 $m^3/s$ (Waite, 1982). Between the Daule and the Babahoyo catchments there is the

Vinces-Quevedo subcatchment (5300 Km$^2$). The river naturally splits part of its flow to the Nuevo River some kilometers upstream Vinces town. Further downstream, the Nuevo River is interconnected with the Abras de Mantequilla (AdM) wetland. This connection takes place in the lower part of the Chojampe subbasin (Fig. II.1-1).

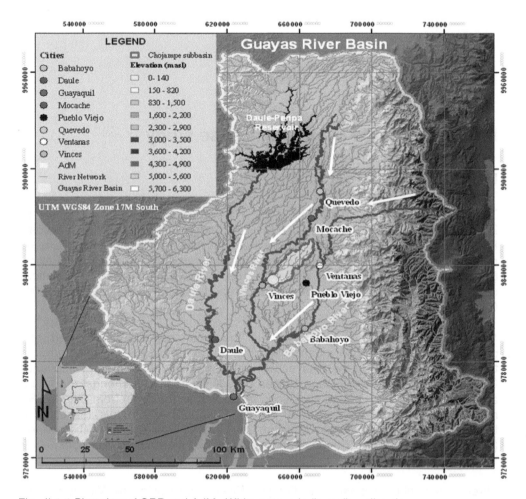

Fig. II.1-1 Plan view of GRB and AdM. White arrows indicate flow directions.

According to the Köppen-Geiger climate classification (Kottek et al., 2006), there are two main climatic zones in GRB. Spatially, around 75% to 80% of the basin belongs to the *Equatorial winter dry zone* (Aw), whereas the northernmost portion (where it rains more) corresponds to the *Equatorial monsoonal type* (Am). From a seasonal perspective, there are only two climatic periods in the basin due to the proximity to the equator. The first one, the rainy, goes from mid-December up to mid-May whereas usually the dry season rules during the rest of the year. Typical intervals of average annual precipitation range from 300 to 4000 mm/year and even beyond that during El Nino, 300% in excess compared to a typical year (Nieto, 2007). A monthly pattern on spatially distributed rainfall is shown in Fig. II.1-2, where the two seasons can be observed.

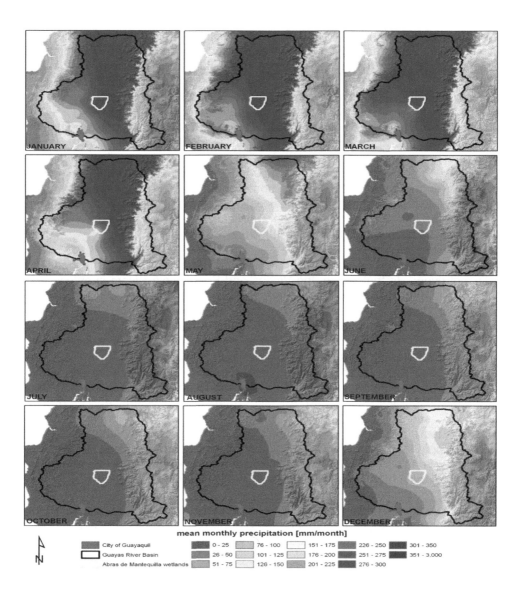

Fig. II.1-2 Spatial variability of monthly rainfall over GRB (Debels et al., 2009).

## II.2   The Abras de Mantequilla wetland system

The Abras de Mantequilla wetland system (AdM) is located in the center of the Los Rios province in Ecuador.   It is a Ramsar site since the year 2000 (BirdLife International, 2012).  AdM entirely fits within the second topographic sub-region of GRB (lowlands).  From a socioeconomic and management point of view, it is the Ramsar site area (Fig. II.2-1) which has been considered by the Ministry of Environment of Ecuador as the wetland domain (56890 Ha).

Hydrographically, a large percentage of the wetland system considered for this study is located inside the Chojampe River catchment (290 Km$^2$) (Fig. II.2-1).  For practical reasons, these wetlands and two extra water bodies outside Chojampe are the only ones to be included in the analysis (described in section IV.5).  Two main reasons support this policy.  Firstly, the few data that is available in the area is mostly located within the aforementioned subbasin.   Secondly, although the available cartography (1980) shows some water bodies outside the Chojampe subbasin, most of them have been desiccated (due to agricultural expansion) in the last decades and are no longer existent in field, being mainly the ones inside the Chojampe subbasin which have currently survived.

Morphologically, AdM is a set of natural swamps and lakes strongly influenced by adjacent rivers.  In that sense, the Vinces and Puebloviejo Rivers overflow part of their waters onto the wetland area during the rainy season.  This is also the case of the Nuevo River and the lower part of the Chojampe Subbasin.  During the dry months, however, an important volume is retained due to natural and man-made embankments.  The wetland contributes to groundwater recharge during the high flood season allowing thus the opposite effect from June to November.  Usually during the high rainy season (March), water depths reach up to 5-8 meters and the wetland branches are beyond 100-400 m wide, particularly near the confluence with the Nuevo River (Fig. II.2-2). An assessment on ecosystem services (Kotze et al., 2008) suggested that the most relevant services AdM provides are the maintenance of biodiversity, cultivated foods and extraction of natural resources, water supply for human use, sediment trapping and erosion control, and streamflow regulation (Fig. II.2-3).  In addition, some stakeholders provided their opinions on the matter (Arias-Hidalgo et al., 2012) (Chapter VI).

In biological terms, inside the wetland area there are still some residuals of a lowland dry forest, some of them inundated during high discharge peaks, as part of the bio-zone of the Gulf of Guayaquil and the Tumbes region.  Around 728 species of flora have been identified and clustered in canopy, emerging and lianas.  These floras are the source of nutrients for the fauna living in the wetland.  Amongst these we have *Crataeva Tapia, Guadua angustifolia, Prosopis juliflora, Capparis angulata and Mutingia* calabura (BirdLife International, 2012).   Moreover, around 14 ictiofauna species have been discovered, classified in 11 families and 5 classes: Mammals, such as the long-tailed otter (*Lontra longicaudis*), birds (more than 100 waterfowl species), reptiles (e.g. *Tortuga mordedora* -Chelydra serpentine), crustaceans (e.g. *Cherax qudricharitnatus*) and fishes, e.g.*"vieja azul"-Aequidens rivulatus* (Prado et al., 2004).

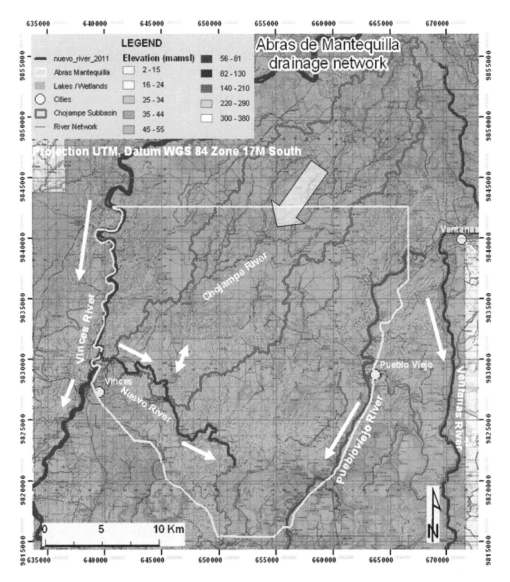

Fig. II.2-1 Drainage network around AdM. Flow directions (white arrows).

Fig. II.2-2 Wet and dry seasons at the AdM (above and below, respectively).

Politically, the AdM Ramsar site comprises part of five municipalities: Vinces, Palenque, Baba, Puebloviejo, and Ventanas (Fig. II.2-4). Since they are relatively close to each other and face similar administrative issues and problems, in late 2008 these municipalities formed with the ones of Mocache, Quinsaloma and Urdaneta an administrative entity known as the *Commonwealth for the Abras de Mantequilla wetland* (http://mancomunidadabras.com.ec/humedal/). This public institution is in charge of natural resources management, and by definition the Abras de Mantequilla. This task is usually carried out in cooperation with SENAGUA (National Ministry of Water), the Los Rios province Council and the Ministry of Environment of Ecuador (Noroña, 2009).

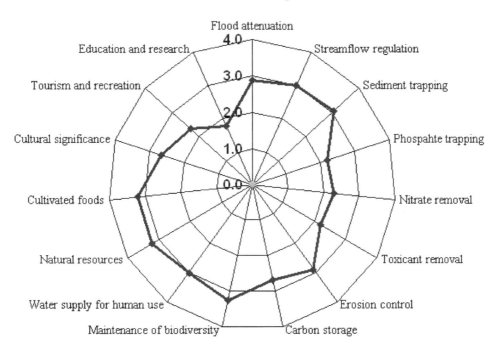

Fig. II.2-3 Ecosystem services evaluation (Kotze et al., 2008) applied on AdM.

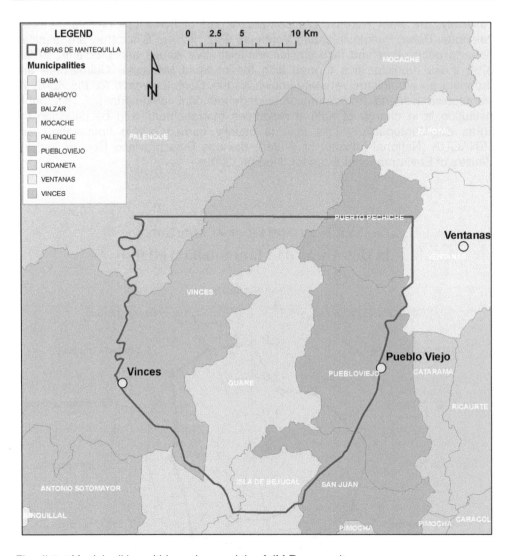

Fig. II.2-4 Municipalities within and around the AdM Ramsar site.

## II.3   DPSIR chains and main pressures on the system

DPSIR stands for Drivers, Pressures, States, Impacts and Responses.   It is a causal chain analysis that has been extensively used in European projects, related often to the Water Framework Directive (WFD) (EC, 2000a).   Drivers are defined as those exogenous / endogenous agents that exert a pressure on the system. Pressures are the variables associated with the drivers, giving a measure of the damage.   State corresponds to the current (or future) situation of the system.

Impacts are the consequences that the drivers cause once they take place. Finally, the responses are the ways how the system reacts either by itself or when human beings intervene (policies / management solutions).

The DPSIR analysis that was carried out during the WETWin project (Zsuffa, 2008; Zsuffa and Cools, 2011) revealed two main influences in the study area, in both basin and wetland scale. The first one was the envisaged infrastructure works by the National Water authority (SENAGUA), namely, the Baba Multipurpose reservoir (Efficacitas, 2006) and the Dauvin project (ACOTECNIC, 2010). These hydraulic projects may introduce remarkable changes in the hydrological pattern of the system. Therefore, it is crucial to evaluate their consequences on the river basin as well as on the wetland.

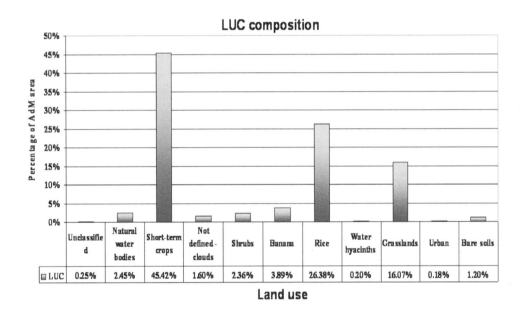

Fig. II.3-1 Land use cover (LUC) in the Abras de Mantequilla wetland area.

The second menace that was identified with the DPSIR chain analysis was the landuse degradation in the wetland area. When this WET-Health assessment (Macfarlane et al., 2008) was implemented on the case study some interesting facts were observed. Although the hydrologic and geomorphologic conditions of the system are currently in an acceptable status with only moderate modifications due to anthropogenic activities, the vegetation status is not. The overall score in the vegetation category was "F" which implies dangerous levels of degradation. A landuse cover (LUC) chart and map of the wetland area can be seen in Fig. II.3-1 & Fig. II.3-2. The massive deforestation in the Ecuadorian coastal region that took place in past decades left its footprints also on the AdM area, where the forest cover is less than 3% of the total. Former forest cover has been replaced by mainly short-term crops, namely rice, maize and beans. In addition, grasslands (for livestock grazing) constitute a remarkable percentage of the area (16%). The intensive use of fertilizers and pesticides (mainly yellow and red label ones) and the piling/burning process (more than 50000 Ha of short-term crop waste/year) may have already causing a toll on the water quality of the water bodies. Hence, incorporating all those factors, a degradation index was evaluated as one of the indicators for the DSS analysis (Chapters V and VI).

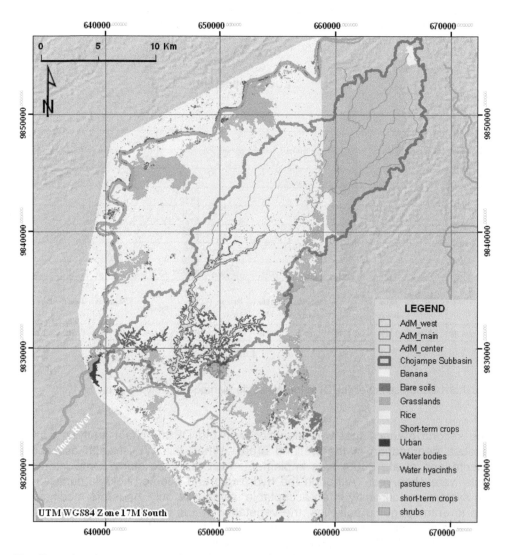

Fig. II.3-2 Landuse cover map (2008) of the AdM wetland.

## II.4   Scenarios

In order to deal with the challenges that system may face in a nearby or mid-term future a set of scenarios and management options have been determined and proposed, respectively.  These had to be in line with the DPSIR chains mentioned before.  In this research, there were two sorts of scenarios, which together gave origin to the *Business As Usual* status (BAU), equivalent to the Management Option 0 (to be described shortly).

- Climate changes;
- The implementation of major infrastructure works, planned by the National Water Authority (The Ministry of Water Resources, SENAGUA).  Important to specify here is that those projects have national priority, therefore, although basically they are human drivers, it was not up to this research to avoid or halt the planning and/or construction should be the case.

### II.4.1   Climatic variations

The Potsdam Climate Institute of Germany (PIK) provided projected rainfall series, based on historical records of maximum and minimum air temperature in the study area.  PIK used the STAtistical Regional model, STAR (Orlowsky et al., 2008).  The hyetographs corresponded to -0.5°C, -1°C, +0.5°C, +1°C and +1.5°C variations of air temperature, throughout a projected time span from 2007 to 2043.  According to local literature, climatic variations for air temperature tend to be positive for the Ecuadorian Coastal region, around 0.0075°C/year in average (Nieto et al., 2002).  Thus this may result in an increment of precipitations, either for a typical year or even during some El Nino events (Vuille et al., 2000).

### II.4.2   The Baba multipurpose project

The Baba dam, whose construction is expected to be finished by late 2013, has various purposes:  irrigation, hydropower generation, flood control and water transfer to the existent Daule-Peripa reservoir (5.5 km$^3$).  The reservoir operative capacity is around 138 hm$^3$ (Efficacitas, 2006) and it is formed at the junction of the Baba and Toachi rivers (Fig. II.4-1).  The Baba reservoir will transfer a maximum of 234 m$^3$/s to the Daule-Peripa dam through a canal 8 km long and the Chaune River during the dry season (Efficacitas, 2006) (Alvarez, 2007).

Since this project includes water retention and diversion, important volume reductions are expected on the lower course of the Vinces River.  In this regard, Efficacitas (2006) suggested an ecological flow value of 10 m$^3$/s that may ensure not only navigability but also sustainability for the living species along the river.  Such a criterion could be extended, via the Nuevo River, to the Abras de Mantequilla wetland.

## II.4.3   The DauVin project

Another initiative of SENAGUA, the DauVin project will profit from a potential excess of water along the Daule river, especially during the dry months. A large area between the Daule, Vinces, Nuevo and Puebloviejo Rivers will be irrigated, helping thus to its socio-economic development. This low zone is traditionally prone to floods during the rainy season and conversely suffers droughts between May and December. Water will be transferred from the Daule River through an artificial canal, sharing water in each river and stream along its path as follows (ACOTECNIC, 2010) (Fig. II.4-2):

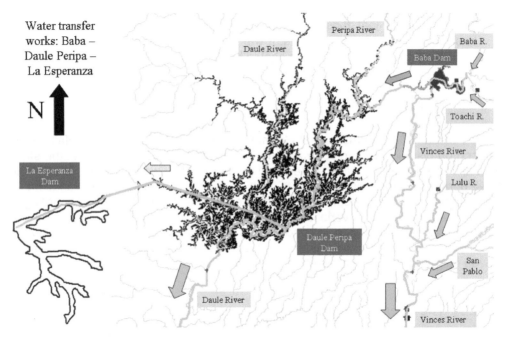

Fig. II.4-1 Water transfers: Daule Peripa to La Esperanza (in operation) and Baba to Daule Peripa (under construction). Other arrows indicate natural flows.

- An initial diversion from the Daule river (96 m3/s).
- A diversion (7.5 m$^3$/s) towards the Estero San Vicente.
- A diversion (63 m$^3$/s) towards Macul River.
- The rest of the main canal (around 25.5 m$^3$/s) will flow along a siphon under the Vinces River, and will enter as an artificial canal to the old path of the Nuevo River where two more minor diversions will take place. The first is towards another canal (Pascuencal) and the latter will go to a natural stream (Estero El Diablo). Finally, the original canal will join the Nuevo River again, 1.7 Km south of the AdM mouth.

- The last canal continues until the Puebloviejo River (6 m$^3$/s) and finally to the Colorado River (out of the study area).

The DauVin project is an initiative of SENAGUA (Water Ministry in Ecuador). Its main objective is to use the water excess along the Daule River, particularly during the shortage months in favour of some drought-prone areas. In order to irrigate extensive zones between the rivers Daule, Vinces, Nuevo and Puebloviejo, an artificial canal heading eastwards from the Daule River will provide water to each of the rivers and small streams along its way (ACOTECNIC, 2010). Since it was not clear in the ACOTECNIC report whether there were implications (positive or negative) on the Nuevo River and the AdM wetland, this thesis included this project as a part of the scenarios to be assessed (Section IV.6).

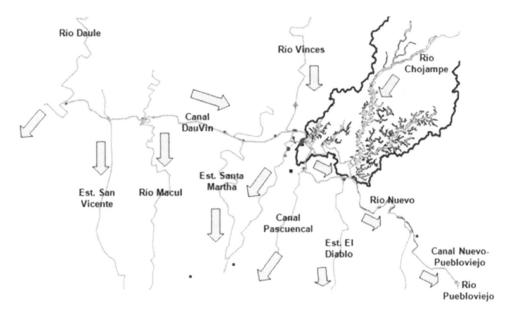

Fig. II.4-2 General schematization of the DauVin project (distribution canals in orange)

## II.5 Management Options & Solutions

The DPSIR causal chains evaluation (Zsuffa and Cools, 2011) and the rest of the baseline results were presented to the Technical Secretariat of the AdM Commonwealth of Municipalities. This governance body collaborated with the ESPOL (Polytechnic University of Coastal Region in Ecuador) staff to devise a set of management options (MO). These MOs focused mainly on minor hydraulic works and landuse improvement as follows:

## II.5.1  Option 0 - *Business As Usual*

As mentioned beforehand, this baseline scenario only considers exogenous drivers that are out of the local management control. It accounts for the combined impact of the introduction of major infrastructure works and climate change. The BAU evaluates what would be the environment behaviour should no management solutions be implemented. Therefore, in order to assess its performance, every MO will be compared against the baseline.

## II.5.2  Option 1 – Minor infrastructure works.

It was suggested to increment the wetland volume thanks to the use of retention gates at the connection point between the main Abras and the Nuevo River. The ground elevation of the natural weir, in average, is 9.80 m.a.s.l., entailing an approximate volume of 8.15 $Hm^3$. If a fixed elevation is kept in 14 m (i.e. gates 4 meters high) during the dry period (June to December), a reasonable water storage for environmental flows might be secured as well as the navigability and the protection of the ichthyologic species.

## II.5.3  Option 2 – Improvement of agricultural practices.

This local-scale alternative adopts a plan of enhancement for short-term crop farmers. Its main objective was the promotion of compost and prohibition of red (extremely toxic) and yellow label (highly toxic) pesticides as well as the reduction of fertilizers, given their worldwide known relation with cancer cases and genetic malformations on children. It was suggested as a policy compromise that 10% of the total landuse surface of the wetland will assume this policy each decade and cumulatively up to the total time span of 4 decades.

## II.5.4  Options 3 & 4 – Crop conversion.

The idea is to substitute short-term crops (e.g. corn) for perennial agroforestry, namely cocoa and fruit trees. This modification in landuse may also contribute to rescue an agricultural tradition because the Los Rios Province was in the XX[th] century the production core of Ecuadorian cocoa. The difference between options 3 and 4 is based on the substitution rate. Option # 3 aims to substitute 10% of the short-term LUC extension per decade, whereas option # 4 increases that rate to 20%. Both have a cumulative effect.

## II.5.5  Option 5 – Ecological corridors

Ecological corridors are planning and implementation units for conservation measures. They are vegetation areas that join two separated ecosystem zones, affected by human intervention. Amongst their targets there are the recovering of degraded zones, the strengthening and enlarging of protected areas and the increment of biodiversity. Knowing that the forest cover does not reach more than 3% of the AdM area (as mentioned earlier), an urgent expansion of these zones is of paramount importance. In this alternative, natural vegetation reforestation is proposed substituting short-term-crops by forest vegetation. The chosen rate was 5% of LUC area per decade. In this regard, this complies with a national campaign promoted by the Ministry of Environment (Ecuador), known as *Socio-bosque* (Lascano, 2009). This program financially encourages those farmers who help reducing the deforestation rates in the country either by preserving the existent natural forests within their properties or reforesting with native species. Therefore, AdM might be an excellent candidate for this initiative.

## II.5.6   Management Solutions

O1 (water storage) and O2 (agricultural practices) were considered as the most important management options and hence they had to be present in every proposed management combination / solution. Reasons for that were:

- It is expected that remarkable improvement in water quantity and water quality is achieved more effectively when water storage is incorporated.
- The enhancement of agricultural practices is currently a State policy; therefore it is advisable to match this national interest with local regulations and practices.

Taking in consideration the priority of the first two alternatives, these MOs were combined into management solutions (MS), in a similar way as they would have been arranged within an integrated management plan:

- MS0: BAU (already including climate changes, Baba Dam & DauVin project).
- MS1: O1 + O2.
- MS2: O1 + O2 + O3.
- MS3: O1 + O2 + O4.
- MS4: O1 + O2 + O3 + O5.
- MS5: O1 + O2 + O4 + O5.

# II.6   Data collection

Data collection for the case study started with topographic information. A Digital Elevation Model (DEM) was built up, based on the latest available survey (1:50000), provided by IGM (Military Geographic Institute of Ecuador). The digitalization work had been already done by Dr. Marc Souris (IRD in France) and published on his website (http://www.rsgis.ait.ac.th/~souris/ecuador.htm), being existent for the whole Republic. On Dr. Souris' webpage there are several available spatial resolutions where the finest is 30m. This dataset was projected in

UTM, datum PSAD56 (Projection for South America 1956), and Zone 17M South; however later on it was re-projected to the WGS 84 datum.

Fig. II.6-1 Available cartography for the Guayas River Basin.

Nevertheless, it was observed that for some flat areas, namely around the Abras de Mantequilla, Dr. Souris' data did give problems for catchment delineation. As an alternative, the Shuttle Radar Topography Mission (SRTM, in resolution 90 m x 90 m, but resampled to 30m) maps were also available. These maps are freely downloadable (http//www.ambiotek.com/srtm). In addition, a 1:10000 scale topography was available for the AdM area. Finally, in order to refine the information in the area where the Chojampe and Nuevo Rivers meet (wetland scale), a local topographic survey was conducted and merged into the existent 1:10000 DEM (Fig. II.6-1).

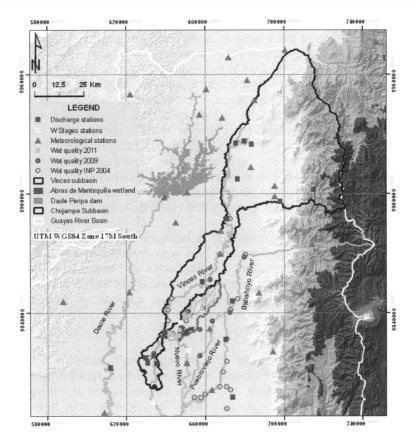

Fig. II.6-2 Main measurement stations around the Vinces subbasin and AdM.

There was a data gathering on the hydro-meteorological stations around the wetland and along the Quevedo-Vinces and Babahoyo-Ventanas-Zapotal river networks, (surrounding the wetland on the west and east, respectively, Fig. II.6-2). To accomplish this task, information was requested from the pertinent institutions. In Ecuador these were SENAGUA (National Ministry of Water), INAMHI (National Institute of Meteorology and Hydrology), and Hidronación (private company in charge of hydropower generation at the Daule-Peripa dam). The time series had mostly a daily resolution (e.g. rainfall, water stages and river discharge). As it is was expected, there were several data gaps. Only a small group of stations reported complete data on rainfall. Other parameters, such as potential evapotranspiration or relative humidity were available in few stations only apart of the fact that their time series had very long data gaps.

Through a recent survey for the WETWin project, it was confirmed a groundwater connection between the Vinces River and the Abras de Mantequilla (Romero et al., 2009). This fieldwork included the measurement of (i) groundwater stages (well sounding); (ii) Characterization of the geological units (vertical electrical sounding); and (iii) Determination of the aquifer's permeability (through a pumping test). The aquifer state beneath the wetland varies from free condition to semi-confined, with an average thickness of 40m. Due to this features, there is a quick water infiltration process. Hydraulic conductivities (rates were estimated around 23 m/d) are in general high, allowing the exploitation, especially for water consumption and irrigation with rates from 10 up to 44 liters/second. This explains the numerous wells observed in the strip between the river and the wetland but also across its surrounding areas (Fig. II.6-3). However, due to budget constraints, this study only covered the mentioned strip and neither the Vinces catchment nor the rest of the AdM area. This is the reason why, for the rainfall-runoff models, only rainfall was considered as direct input data, being evapotranspiration, interception, and groundwater parameters indirectly estimated (section IV.3).

In other cases, data were discontinuous because some measurement stations were not operational for long periods. This was detected at some river stations, where gaps were found throughout the discharge and water stage hydrographs. Since some of these spots were needed as boundary conditions for a river routing model, and it was not easy to apply traditional approaches (e.g. nearby river stations), a simple filter-based technique was used to estimate the existent gaps (Chapter III).

Regional spatial data were used for the catchment scale (Quevedo-Vinces system). This included available maps of landuse and soil types in scale 1:250000 (CEDEGE, 2002), key information to compute surface runoff at the outlet (section IV.3). In this regard, remotely sensed data (RS) were another helpful resource taking advantage of the already displayed spatial distribution, despite the common constraint of cloudiness over the study area. For landuse cover, an image from 2008 was obtained thanks to CLIRSEN (remote sensing public institution in Ecuador).

Data were even scarcer in the water quality domain. This is because few studies have been conducted in the wetland area (Prado et al., 2004). Mainly due to budget regulations, the National Fishery Institute (INP) collects data only when there are suspicious cases of fish mortality along the rivers or some specific complaint from local fishermen. However, some information on nitrates, nitrites and phosphates was collected in August 2009 and February 2011 along the main branches of Abras de Mantequilla by ESPOL University being that the basis for a further simple mass balance (section IV.6). Finally, data were reorganized in a geo-database using ArcMap 9.3 (Fig. II.6-4). Diverse *Feature datasets* were created when files shared a theme, zone, scale or a specific attribute. Such order was followed for both the vector and the raster files and for the time series (originally in .xls format). Metadata information was filled for every file, using the ISO 19139 format (Fig. II.6-5). Part of this database can be seen at the UNESCO-IHE Spatial Data Infrastructure portal (SDI) (http://sditest.unesco-ihe.org:8080/geo network).

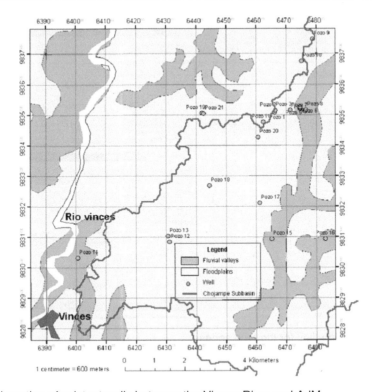

Fig. II.6-3 Location of existent wells between the Vinces River and AdM.

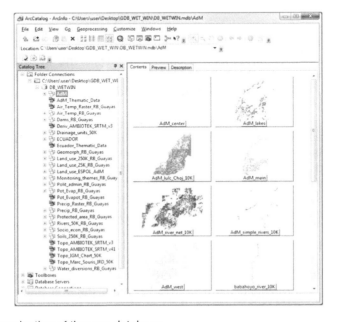

Fig. II.6-4  Organization of the geo-database.

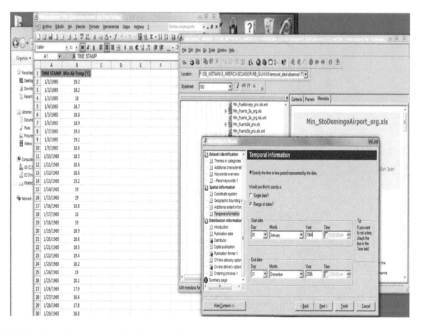

Fig. II.6-5 Metadata files for the time series.


## II.7   Concluding remarks

The Abras de Mantequilla, a Ramsar site has been chosen as the case study of this research.  This wetland system is in the middle of the Guayas River Basin, one of the main riverine ecosystems in Ecuador.  Despite the overall low degree of human intervention, the DPSIR analysis established two main pressure drivers on the system:  the major infrastructure works at a basin scale, planned by SENAGUA and the landuse degradation.

In order to assess the potential influences of these events, a composed scenario, the Business As Usual and a set of management solutions at a local scale have been envisaged and described.  The BAU scenario included the potential climatic variations as well as the modifications that the Baba and DauVin projects may entail on the system.  On the other hand, the management solutions mainly covered water retention and different degrees of landuse changes.

The recurrent lack of data was a major obstacle to the system characterization, the model building process and ultimately, the decision making process.  Facing this drawback and for the streamflow time series scattered across the study area, a methodology to characterize them and to estimate gaps is presented shortly in Chapter III.  With analogous purpose, the use of rainfall satellite data is explored as a suitable complement to the ground raingauges (Chapter IV).

*Chapter Three*

# III  A simple pattern simulation in daily streamflow series*

## III.1 Introduction

Streamflow time series are of utmost importance for a catchment-hydrological analysis and in general to water resources management. Nonetheless, continuous or reliable data are not always available everywhere. Pattern characterization as well as the estimation of incomplete data in time series are often a challenge for many scientists because they are not easy tasks in engineering. Specially in the early stages of water management, a sound characterization of river discharges series and their seasonal variations is extremely important (Black, 1996; Simonovic, 2009).

The analysis key parameters in flow time series, such as high-flow frequencies, magnitude, and rate of change have led to a major advance in the identification of streamflow variations and regional patterns, especially in the field of Ecohydraulics (Poff and Ward 1989; Richter, Baumgartner et al. 1996)  In other researches carrying out a sensitivity analysis, the most relevant flow parameters have been used to characterize the flow patterns in a group of ungauged catchments (Sanborn and Bledsoe, 2006)  Furthermore, from a spatial perspective, strong correlations have been found between flow features and catchment properties for dam impact assessments (Matteau et al., 2009) and spatial classification of streams depending on flow regimes clusters has been recently reported (Moliere et al., 2009).

---

* Chapter based on: Arias-Hidalgo, M., Villa-Cox, G., van Griensven, A., Mynett A. E., (2012), "A simple pattern simulation in daily streamflow series", Journal of Hydrological Sciences (in revision)

One of the possible further applications of streamflow series characterization is the synthetic pattern replication for data gaps. Due to a lower number of seasons, i.e. temporal variations, tropical rivers are less difficult to characterize than the non-tropical ones (Dettinger and Diaz, 2000). And it is precisely in tropical zones where the lack of information is a serious problem. Usually gaps are originated due to the lack of maintenance of the measurement devices, poor reading performance or absence of serious monitoring campaigns or the *culture of consistent measurement taking*. Notwithstanding the cause, missing data constitute a major barrier to understand what is happening in a particular catchment location and of course to build hind/forecasting models. In this regard, a sound series pattern characterization could be very helpful. Starting from simple deterministic techniques up to sophisticated stochastic tools, several attempts have been made to study the trends of streamflow time series, categorize discharge hydrographs with coarse temporal resolution using Fourier Series, estimate potential gaps when required and apply these estimations to hydrologic analyses as reported in previous years (Yevjevich 1972; Kottegota 1980; Kahya and Dracup 1993; Aksoy and Bayazit 2000; Srinivas and Srinivasan 2005).

Either to characterize patterns or to tackle long-gaps, black and grey-box methods seem to be promising alternatives. One example is the Artificial Neural Networks (ANN) and other data-driven methods in hydrology and Environmental Hydroinformatics (Minns and Hall 1996; Mynett 2002; Mynett 2004; Ilunga and Stephenson 2005; Wang, Li et al. 2005; Mynett 2008; Besaw, Rizzo et al. 2010). These methods have been extensively implemented in runoff simulation, dealing with the non-linearity of hydrological processes. Nonetheless, it must be noted that the performance of neural networks and in general black-box techniques, without physically-based knowledge, is strongly dependent not only on the correct selection of the network architecture but also the amount and quality of the available data (Price and Vojinovic, 2011; Wang et al., 2005).

Wavelet analysis is nowadays another interesting option to study observed data and estimate missing series. As shown recently, Haar wavelets and signal processing in general with normally distributed data (Bloomfield, 2000; Wang et al., 2011) can incorporate key statistical features present in data (e.g. mean, standard deviation) and also their variability (Coulibaly and Burn, 2004; Smith et al., 1998). Furthermore, discrete wavelet transforms have been employed with river flow regression models to achieve acceptable forecasting performances (Küçük and Agiralioglu, 2006). Alongside these developments, there is still space for exploring more wavelet-based techniques, such as filters in order to study patterns on river flow time series and perhaps estimate missing data, when necessary. It would be desirable that the pattern characterization technique incorporated not only statistical features and information gain from the streamflow series (Jonsdottir et al., 2008; Pan et al., 2012) but also some physical knowledge such as the influence of climate seasons.

In this regard, it is proposed the use of the so-called Hodrick-Prescott filter (HPF) (Hodrick and Prescott, 1997). Widely adopted in Econometrics, the HPF extracts the statistical characteristics of trade cycles (Schlicht, 2004). The aim of this research is to find a simple, yet operational method that can find series patterns and identify recurrent changes in the hydrograph for a *training* period and, as an application, can estimate the signal pattern during a gap in a relatively easy way. This procedure takes into account the seasonal regimes, as not many studies have yet focused on. Moreover, the usefulness of this methodology can be more appreciated when long periods are to be intervened (e.g. months in hydrological time series). Another probable application can be when there are no nearby river flow stations with correlated data during a particular period, or when there is no available rainfall time series in the time steps prior to the discharge one wants to know.

## III.2 The Hodrick-Prescott filter

### III.2.1 General Equations

The Hodrick-Prescott (Hodrick and Prescott, 1997) is a band-pass filter (de-trending method) that decomposes the original signal $Y_t$ into two elements: the trend, also known as long-term component, running average, long-run or tendency (Phillips, 2010; Walker, 1999), $m_t$; and the noise, a.k.a. short-term, fluctuation or cyclical component, $c_t$. HPF has been extensively used in Macroeconomics (del Río, 1999; King and Rebelo, 1999; Schlicht, 2004). Eq. III-1 illustrates, thus, the two aforementioned elements of the series (Fig. III.2-1):

$$Y_t = m_t + c_t$$

Eq. III-1

The main problem that HPF aims to solve is a minimization of a likelihood function (LF), as follows (Razzak, 1997):

$$\min_{\{m_t\},\{c_t\}} LF = \sum_{t=1}^{T} c_t^2 + \lambda \sum_{t=1}^{T} (\Delta^2 m_t)^2$$

Eq. III-2

In Eq. III-2, the summation of the squared noises is a measure of the goodness of fit whereas the second part includes a penalization to the non-linearity by a smoothing factor, $\lambda$. If $\lambda$ is too high (or approaching infinitely) the filter becomes linear (only depending on the trend). Conversely, when $\lambda = 0$, it entirely depends on the noise; $\Delta = (1-B)$ is the standard differencing operator; and, B the backshift element (Box et al., 2008). Noteworthy is that $c_t$ (the noise) is assumed to have a normal distribution with $\mu = 0$ and known variance, $\sigma^2$ (King and Rebelo, 1999).

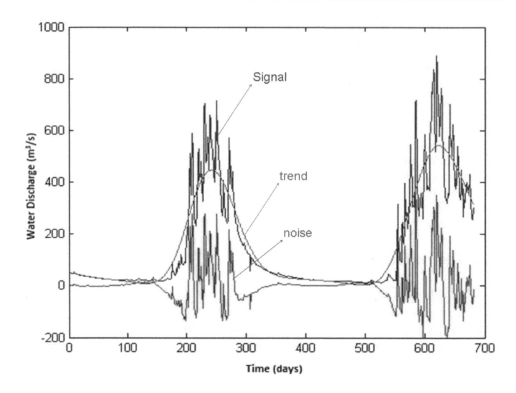

Fig. III.2-1 Discharge hydrograph decomposition using the HPF.

The trend $m_t$ is computed by Eq. III-3 (Danthine and Girardin, 1989), during the training (sampling) period, as a function of the original signal:

$$m_t = (I + \lambda KK')^{-1} Y_t$$

Eq. III-3

where the K matrix is defined in the following way:

$$K_{ij} = \begin{cases} 1, & if \ \ i = j \ \ or \ \ i = j+2 \\ -2, & if \ \ i = j+1 \\ 0, & otherwise \end{cases}$$

Eq. III-4

$\lambda$ is estimated depending on the temporal scale: yearly, monthly; sometimes assumed as equal to 1600 (Ravn and Uhlig, 2002). For daily streamflow series, $\lambda$ was conceived by means of the *1/2 gain rule* (del Río, 1999). Del Río reported that the long-term component's gain, $v^m$ and the short-term component's gain, $v^c$ produced by the HPF in terms of λ and the radian frequency ω of the resulting cycle are given by:

$$v^m(\omega, \lambda) = \frac{1}{1 + 4\lambda(1 - \cos \omega)^2}$$

Eq. III-5

$$v^c(\omega, \lambda) = 1 - v^m(\omega, \lambda)$$

Eq. III-6

By setting the filter's gain to 0.5, the following relation is obtained:

$$\lambda = f(\omega) = \frac{1}{4(1 - \cos \omega)^2}$$

Eq. III-7

Moreover, there is a convenient relation between ω and the period $\tau$ of the cycles produced by the HPF, i.e. $\omega = 2\pi / \tau$. Combining this expression with Eq. III-7, λ is expressed as a function of $\tau$:

$$\lambda = f(\tau) = \frac{1}{4(1 - \cos \frac{2\pi}{\tau})^2}$$

Eq. III-8

Finally, Eq. III-8 may be a suitable way to determine λ. Through the direct observation of the discharge hydrographs at each involved monitoring station, it is possible to obtain the average duration of the rainy season in the area (in days). This might be a reasonable estimation for $\tau$. By means of this parameter, more information about typical behaviours or patterns of the long-run trend component ($m_t$) and the short-term noise ($c_t$) can be learnt.

## III.2.2  The behaviour of the short-term component

A classical approach when applying the HPF to a time series would be to assume that the generated short-term component is distributed as $c_t \sim N(0, \sigma^2)$. In other words, $c_t$ would be considered *white noise* with a known variance. However, this initial assumption was discarded because it was observed that the magnitude of the noise variance was not constant throughout the series. Actually, as seen in Chapter II, in the Guayas River Basin there are two clearly different seasons: the wet (from late December to late April or early May) and the dry for the rest of the year.

Table III.2-1 Historical statistical features of the river stations in the Guayas River Basin involved in this research.

| Streamflow station | River | Time span | Mean $(m^3/s)$ | Std Dev $(m^3/s)$ | Max $(m^3/s)$ | $\tau$ (days) | Thresh old 1 | Thresh old 2 |
|---|---|---|---|---|---|---|---|---|
| Baba project | Baba | 1964-2006 | 107 | 115 | 1148 | 161 | 60 | 72 |
| Daule at La Capilla | Daule | 1982-2007 | 287 | 369 | 1953 | 131 | 60 | 72 |
| Mastrantal DJ Pula | Mastrantal | 1971-1994 | 50 | 50 | 223 | 143 | 8 | 10 |
| Nuevo DD Vinces | Nuevo | 1972-2004 | 34 | 50 | 530 | 108 | 40 | 48 |
| Pula at Palizada | Pula | 1975-2006 | 77 | 68 | 277 | 163 | 25 | 30 |
| Quevedo en Quevedo | Quevedo | 1962-2006 | 220 | 255 | 2247 | 125 | 120 | 144 |
| Toachi AJ Baba | Toachi | 1984-2007 | 36 | 74 | 983 | 122 | 24 | 29 |
| Vernaza DD Vinces | Vernaza | 1971-2008 | 88 | 98 | 366 | 177 | 30 | 36 |
| Vinces en Vinces | Vinces | 1965-2006 | 224 | 232 | 955 | 163 | 200 | 240 |
| Vinces Hcda. Casa | Vinces | 1984-1991 | 163 | 205 | 1354 | 149 | 166 | 200 |
| Zapotal at Catarama | Babahoyo | 1971-2007 | 173 | 202 | 930 | 166 | 130 | 156 |
| Zapotal at Lechugal | Babahoyo | 1984-1991 | 163 | 205 | 1354 | 117 | 60 | 72 |

Two alternatives are proposed to represent the two seasons based on the assumption that $c_t$ exhibits different patterns throughout a year. These patterns are associated with the long term component $m_t$. Both approaches rely upon the observation of $m_t$ to estimate a threshold value that may a distinction between dry season and wet season. Observing the available series throughout the time span, some threshold values were determined, for each station (Table III.2-1). Threshold 1 marked the beginning of the rainy season; conversely, the value limit 2 indicated the end of it. In general, it was found that the threshold 2 was around 20% higher than the 1. The time indices of dry and wet regimes were denoted $t \in S_1$ and $t \in S_2$, respectively. Thus, an empirical seasonal classification was achieved.

## III.2.3  The Block Homoskedastic approach

In this case, it is assumed that there is one standard deviation associated with each one of the seasons. The suggested behaviour of $c_t$ under this approach is:

$$c_t \sim N(0, \ \sigma_t) \ ; \ \sigma_t = \begin{cases} \sigma_1 \ \ if \ t \in S_1 \\ \sigma_2 \ \ if \ t \in S_2 \end{cases}$$

Eq. III-9

A key point here is that both sets of indices in $S_1$ and $S_2$ can be obtained deterministically. This is to maintain the independence of the sample conformation and to allow the estimation of parameters $\sigma_1$ and $\sigma_2$ by means of Maximum Likelihood Estimation (MLE). The likelihood function for Eq. III-9 is:

$$L = \prod_{t=1}^{T} \frac{1}{\sigma_t \sqrt{2\pi}} \exp\left(-\frac{c_t^2}{2\sigma_t^2}\right)$$

Eq. III-10

$$\ln L = -\frac{1}{2} T \ln \ (2\pi) - \Sigma_{t \in S_1}\left[\ln \ \sigma_1 + \frac{c_t^2}{2\sigma_1^2}\right] - \Sigma_{t \in S_2}\left[\ln \ \sigma_2 + \frac{c_t^2}{2\sigma_2^2}\right]$$

Eq. III-11

where $T = T_1 + T_2$; $T_1$ and $T_2$ stand for the total number of elements in $S_1$ and $S_2$ respectively. From Eq. III-11 the first order conditions are:

$$\frac{\delta \ln L}{\delta \ \sigma_1} = \Sigma_{t \in S_1}\left[-\frac{1}{\sigma_1} + \frac{c_t^2}{\sigma_1^3}\right] = 0$$

Eq. III-12

$$\frac{\delta \ln L}{\delta \ \sigma_2} = \Sigma_{t \in S_2}\left[-\frac{1}{\sigma_2} + \frac{c_t^2}{\sigma_2^3}\right] = 0$$

Eq. III-13

By solving Eq. III-12 and Eq. III-13 it is possible to demonstrate that:

$$\hat{\sigma}_1 = \sqrt{\frac{\Sigma_{t \in S_1} c_t^2}{T_1}}$$

Eq. III-14

and:

$$\hat{\sigma}_2 = \sqrt{\frac{\sum_{t \in S_2} c_t^2}{T_2}}$$

Eq. III-15

## III.2.4   The Heteroskedastic approach

This second proposed approach is slightly more complex than the first, but allows more flexibility to simulate the behaviour of $c_t$ :

$$c_t = \sqrt{(\alpha + \gamma_t \beta)|m_t|^{\gamma_t}} . v_t \; ; \; \gamma_t = \begin{cases} 0 \; if \; t \in S_1 \\ 1 \; if \; t \in S_2 \end{cases}$$

Eq. III-16

where $v_t \sim N(0,1)$, $E(c_t | m_t) = 0$ and $V(c_t | m_t) = (\alpha + \gamma_t \beta)|m_t|^{\gamma_t}$. This implies that the variance of $\hat{\varepsilon}_t$ behaves in similar manner as in the first approach since $V(c_t | m_t) = \alpha$ if $t \in S_1$. On the other hand, when $t \in S_2$, the variance is proportional to $m_t$ because $V(c_t | m_t) = (\alpha + \beta)m_t$. In other words, for the dry regime the values of $m_t$ do not show sufficient variation to affect the behaviour of the variance of $c_t$; conversely, during the wet regime, the variance of $c_t$ is proportional to $m_t$ due to significant variations of the trend. The likelihood function for Eq. III-16 is as follows (Greene, 2008):

$$L = \prod_{t=1}^{T} \frac{1}{\sqrt{2\pi(\alpha + \gamma_t \beta)|m_t|^{\gamma_t}}} \exp\left( -\frac{c_t^2}{2(\alpha + \gamma_t \beta)|m_t|^{\gamma_t}} \right)$$

Eq. III-17

$$\ln L = -\frac{1}{2}T \ln (2\pi) - \sum_{t=1}^{T} \left[ \frac{1}{2}\ln (\alpha + \gamma_t \beta) - \frac{\gamma_t}{2} \ln|m_t| - \frac{c_t^2}{2(\alpha + \gamma_t \beta)|m_t|^{\gamma_t}} \right]$$

Eq. III-18

$$\ln L = -\frac{1}{2} T \ln (2\pi) - \frac{1}{2} \Sigma_{t \in S_1} \left[ \ln \left( \alpha + \frac{c_t^2}{\alpha} \right) \right] - \frac{1}{2} \Sigma_{t \in S_2} \left[ \ln (\alpha + \beta) |m_t| + \frac{c_t^2}{(\alpha + \beta)|m_t|} \right]$$

Eq. III-19

From Eq. III-19 the first order conditions are:

$$\frac{\delta \ln L}{\delta \alpha} = \Sigma_{t=1}^T \left[ -\frac{1}{2(\alpha + \gamma_t \beta)} + \frac{c_t^2}{2(\alpha + \gamma_t \beta)^2 |m_t|^{\gamma_t}} \right] = 0$$

Eq. III-20

$$\frac{\delta \ln L}{\delta \beta} = \Sigma_{t=1}^T \left[ -\frac{\gamma_t}{2(\alpha + \gamma_t \beta)} + \frac{\gamma_t c_t^2}{2(\alpha + \gamma_t \beta)^2 |m_t|^{\gamma_t}} \right] = 0$$

Eq. III-21

To solve Eq. III-20 and Eq. III-21 is not as straightforward as it was for the Block Homoskedastic approach. Nevertheless, these expressions can still be reformulated in the following way:

$$\Sigma_{t \in S_1} \left[ \frac{1}{\alpha} + \frac{c_t^2}{\alpha^2} \right] + \Sigma_{t \in S_2} \left[ -\frac{1}{\alpha + \beta} + \frac{c_t^2}{(\alpha + \beta)^2 |m_t|} \right] = 0$$

Eq. III-22

since $$\Sigma_{t \in S_2} \left[ -\frac{1}{\alpha + \beta} + \frac{c_t^2}{(\alpha + \beta)^2 |m_t|} \right] = 0$$

Eq. III-23

hence $\Sigma_{t \in S_1} \left[ \dfrac{1}{\alpha} + \dfrac{c_t^2}{\alpha^2} \right] = 0$, and $\alpha$ is estimated as:

$$\hat{\alpha} = \frac{1}{T_1} \Sigma_{t \in S_1} c_t^2$$

Eq. III-24

Finally, $\beta$ is estimated inserting Eq. III-24 into Eq. III-23:

$$-\frac{T_2}{\alpha+\beta}+\frac{1}{(\alpha+\beta)^2}\sum_{t\in S_2}\frac{c_t^2}{|m_t|}=0$$

$$\hat{\beta}=\frac{1}{T_2}\sum_{t\in S_2}\frac{c_t^2}{|m_t|}-\frac{1}{T_1}\sum_{t\in S_1}c_t^2$$

Eq. III-25

## III.3  A further application: estimation of gaps in time series

### III.3.1  Fourier series

Fourier analysis is a technique that can be use to project the trend that was computed during the training length. Consider a given signal Q with a given periodicity T. Q then can be approximated as the summation of an infinite number of sinusoidal functions. Important to have in mind is that Fourier analysis can be applicable when the series have a constant sampling interval $\Delta t$ (daily in the present case). Being f the original sinusoidal signal frequency and f' the frequency of another sinusoidal function, both series are said to be aliased (i.e. look quite similar to each other) if (f-f') is multiple of $1/\Delta t$. The Euler-Fourier infinite sum for the Fourier analysis is as follows:

$$f(x)=\frac{a_o}{2}+\sum_{m=1}^{\infty}(a_m\cos\frac{m\,\pi x}{L}+b_m\sin\frac{m\,\pi x}{L})$$

Eq. III-26

where m is the order of the summation that in this study is finite with m=8; L = m*T/2 and T is the fundamental period for each of these sinusoidal and co-sinusoidal functions. Given a positive integer $n$, and taking advantage of the mutual orthogonality of these functions when m=n, the regression coefficients $a_m$ and $b_m$ can be computed as:

$$a_n=\frac{1}{L}\int_{-L}^{L}f(x)\cos\frac{n\,\pi x}{L}\,dx,\qquad n=0,1,2,...M$$

Eq. III-27

$$b_n = \frac{1}{L} \int_{-L}^{L} f(x) \sin \frac{n\,\pi x}{L}\, dx, \qquad n = 1, 2, 3, ....M$$

Eq. III-28

## III.3.2  Selection of the *best* alternative

The Akaike Information Criterion (AIC) (Akaike, 1974) is a measure of the goodness of fit of an estimated statistical model. It is grounded in the concept of entropy, offering a relative measure of the information that is lost when a given model is used to describe reality.  This criterion also describes the tradeoff between bias and variance in a model construction, or loosely speaking that of accuracy and complexity of the model.  AIC is not a test of a model in the sense of hypothesis testing, rather it is a test between models: a tool for model selection. Given a data set, several competing models may be ranked according to their AIC, with the one having the lowest AIC being the *best*.  In the present study, of course, this criterion was applied to the Block Homoskedastic and the Heteroskedastic alternatives.  Eq. III-29 shows the AIC, which is more formally known as the Akaike's Bayesian Information Criterion or ABIC (Akaike, 1980):

$$ABIC_m = 2 * \ln(T_m) - 2 * L_m$$

Eq. III-29

where for a given model m:
T is the total number of elements = $T_1 + T_2$, $T_1$ for $S_1$ and $T_2$ for $S_2$;
$L_m$ is the natural logarithm of the likelihood function, previously defined in Eq. III-11 and Eq. III-19 for each model approach.

## III.3.3  Computational setup

The entire methodology was constructed building scripts in MATLAB$^{TM}$.  The flowchart summarizing the entire process can be seen in Fig. III.3-1. Initially, the river hydrograph was divided between the tendency and the noise using the HPF. A total of 12 river stations across the Guayas River Basin have been assessed by this methodology (Fig. III.3-2).  Some of these stations were afterwards part of boundary conditions for a hydrodynamic simulation in the study area (section IV.5). Each station has its own $\tau$ value, in days (Table III.2-1) as well as the corresponding time span where gaps with different lengths were artificially prepared.  For all cases, the streamflow data had a daily resolution.  To determine whether a certain time step belonged to the dry or wet period, some observed thresholds in the series were adopted (as explained before).

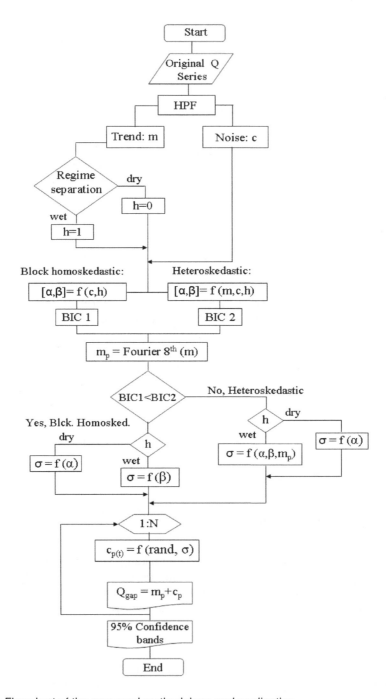

Fig. III.3-1 Flowchart of the proposed methodology and application.

Furthermore, the trend was projected throughout the gap ($m_p$) using the $8^{th}$ order Fourier series.    Then, the maximum likelihood estimators $\hat{\alpha}$ and $\hat{\beta}$ were calculated.    These and the Akaike Information Criterion were computed for each approach: block homoskedastic and heteroskedastic.    The variant with the smaller BIC was the chosen one.    This drove how the standard deviation was computed, as in Eq. III-9 and Eq. III-16.

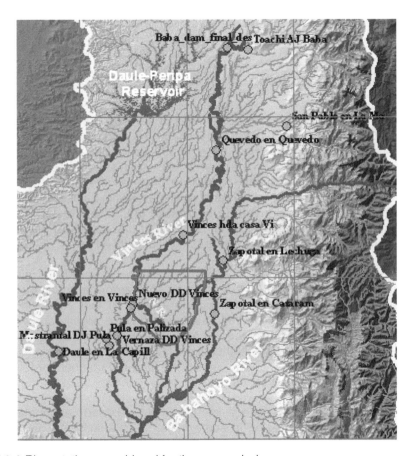

Fig. III.3-2 River stations considered for the gap analysis.

The final stage was based upon a loop where the projected noise ($c_p$) depended on the stochastic error.    This error introduced a random generator that affected the already computed standard deviation being 2500 the total number N of Monte-Carlo simulations.    Later on, the estimated series were generated summing up the projected trend ($m_p$) and the projected noise ($c_p$).    Ultimately, not only a realization but also the 95% confidence interval bands were computed for each estimated gap.    In general, the simulations were carried out having in mind that a potential gap would be within one of the following four intervals:

a) Wet season;
b) Dry season;
c) Transition from late wet to early dry season;
d) Transition from late dry to early wet season;

The gap-estimation application in this research found a justification when facing the problem of large data deficits throughout those river discharge time series. It was not always possible to use regression between the gauging points of the same river. Moreover, the spatial data density across the basin is very low (less than 20 stations with reliable time series data).

## III.4  Results & Discussion

The Hodrick-Prescott filter was tested in series of several durations, ranging from 2 to 6 years. On the other hand, the application for gap extrapolation was carried out designing artificial gaps ranging from 60 up to 300 days long. Because the present application was not aimed to detect or predict peaks and because it was a stochastic approach, any resultant realization between the bands could not be fairly compared with the original streamflow series using traditional deterministic criteria such as Root Mean Squared Error (RMSE) or the Nash-Sutcliffe coefficient (NSC) (Nash J.E. and Sutcliffe J.V., 1970). Rather, the statistical features of the bands and the available series were contrasted (Table III.4-1).

In general, the Hodrick-Prescott filter decomposed very well the original series during the training period. Fig. III.4-1 shows the performance at the *Daule at Capilla* station. In this example a sampling interval of two years showed two different behaviours. First, year 1986 seemed to have a slightly prolonged rainy season going as far as early June; whereas 1987 showed a typical duration (January to early May) but with higher flows. Despite these dissimilarities and some spikes (e.g. July 1987), in both cases the long-run component followed the pattern dictated by the original series.

Due to the phase-shifting effect of the Fourier-based extrapolation, the performance of tool was affected in some trials. This shift was detected after analyzing the first, second and third derivatives of the extrapolated trend. Notwithstanding this problem and for those gap applications during the wet season (Dec-May) it was observed that the pattern was similar to the one of the measured series. A visualization example was taken from the *Nuevo DD Vinces* river station at beginning of the Nuevo River (Fig. III.4-2). In the Ecuadorian lowlands, this period usually shows the highest flows. For the example in mention, the trend's high period was found around early March which was consistent with the pattern at that station. Conversely, it was seen that the method overestimated the observations close to the minimum and up to percentile 25. Thereafter the bands comprised well the rest of statistical features of the observations, including the mean value (Table III.4-1).

Table III.4-1 Summary of statistical features for the gap analysis application.

| Station / Statistics | Percentile | Lower band 2% | Observed | Upper band 97% |
|---|---|---|---|---|
| | Min | 39 | 0 | 101 |
| | 25 | 49 | 23 | 176 |
| | 50 | 61 | 66 | 212 |
| **Nuevo DD Vinces 01** | 75 | 73 | 128 | 237 |
| | Max | 83 | 190 | 250 |
| | Mean | 61 | 78 | 204 |
| | StDev | 13 | 58 | 42 |
| | Min | 1 | 25 | 46 |
| | 25 | 2 | 29 | 68 |
| | 50 | 3 | 32 | 99 |
| **Quevedo at Quevedo 00** | 75 | 4 | 34 | 106 |
| | Max | 6 | 54 | 121 |
| | Mean | 3 | 32 | 90 |
| | StDev | 1 | 5 | 24 |
| | Min | 35 | 23 | 351 |
| | 25 | 68 | 206 | 406 |
| | 50 | 293 | 352 | 811 |
| **Zapotal at Lechugal 03** | 75 | 477 | 584 | 1095 |
| | Max | 587 | 1536 | 1185 |
| | Mean | 286 | 421 | 764 |
| | StDev | 186 | 298 | 319 |
| | Min | 1 | 15 | 62 |
| | 25 | 5 | 23 | 141 |
| | 50 | 14 | 39 | 264 |
| **Daule at La Capilla 00** | 75 | 108 | 269 | 679 |
| | Max | 221 | 905 | 964 |
| | Mean | 59 | 199 | 404 |
| | StDev | 72 | 266 | 308 |

During the dry season, the general observed course for the estimated time series was found to be close to the real ones. In fact, low streamflow magnitudes, typical of this period, seem to support this finding. This was the case of the gap between September 9th and December 9th, 1970 for the *Quevedo at Quevedo* station on the Vinces River (Fig. III.4-3). The differences between the measured discharges (during the gap) and the projected trend could be linked probably to the model learning process during the sampling period (approximately 2 years) and the aforementioned phase shift. Anyway, a comparison of statistical features of a random realization and its 95% confidence bands vs. the observed series is shown in Table III.4-1. For every property, the real (observed) series were within the simulated ranges. Moreover, according to local experiences in data measuring campaigns during this period, the observed data may not truly imply a *stagnant* behaviour, granting thus some more reliability to the estimated bands.

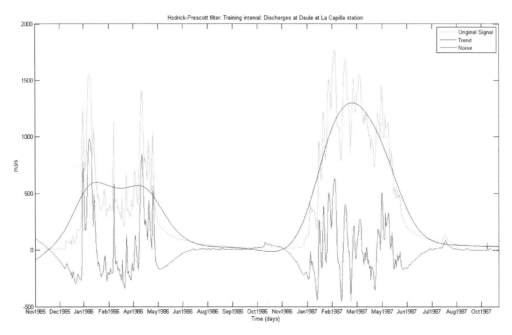

Fig. III.4-1 HPF during the training period. *Daule at Capilla* station

Furthermore, it was found that the trends of the reconstructed and original series were relatively close during the transition period between the high wet and the early dry season. The visual example was taken from the *Zapotal at Lechugal* station on the Babahoyo River where the sampling period lasted for 730 days and the gap was from March 4th to May 4th, 2006 (Fig. III.4-4). Noteworthy is that the two discharge peaks of March 17th and April 1st, 2006 were by far out of the 95% confidence intervals. This was corroborated when observing the statistical properties in Table III.4-1, where only the minima and maxima from the flow observations were beyond the projected ranges. Nevertheless, it is important to highlight that the method did register the season shift to the dry period (around March 29th) as the bands turned narrower thenceforth.

In a similar way the application attempted fairly well to simulate the variation from the dry period (around November) to the wet season (late January). An illustrative case was the *Daule at La Capilla* station (Fig. III.4-5), where the sampling period covered 2 years and the gap lasted 120 days (Oct 31st, 1987 to Feb 28th, 1988). The rising turning point was around mid-December, being in line with the usual behaviour of the river series in that particular region. The measured values were found to be between the 95% confidence bands for every computed statistical parameter (Table III.4-1).

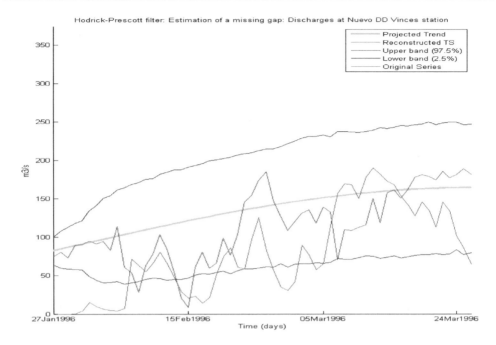

Fig. III.4-2 Gap estimation during the wet season. *Nuevo DD Vinces* station.

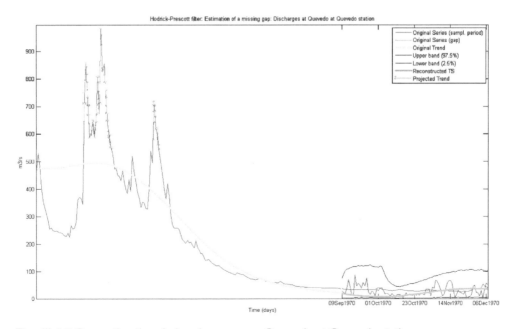

Fig. III.4-3 Gap estimation during dry season. *Quevedo at Quevedo* station.

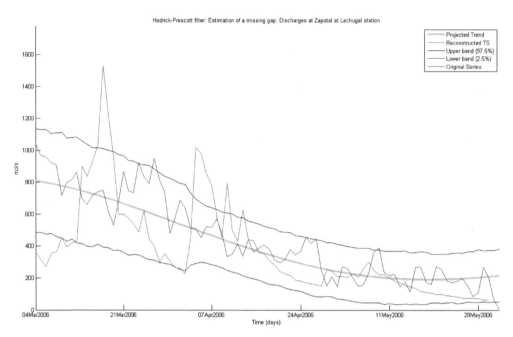

Fig. III.4-4 Gap estimation during the transition period from wet to dry season. *Zapotal at Lechugal* station.

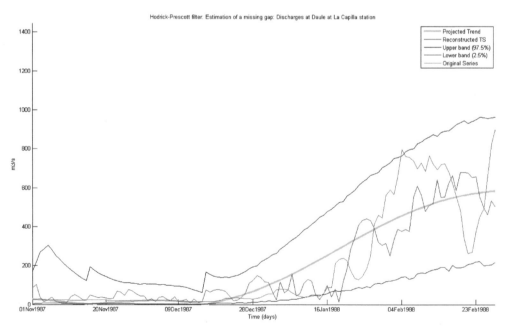

Fig. III.4-5 Gap estimation during the transition period from dry to wet season. *Daule at Capilla* station.

Finally, in an example on the same synthetic gap for the wet season in the *NuevoDDVinces* gauging spot, a linear regression model based on data from the *Quevedo at Quevedo* station was contrasted against the proposed technique. As expected, the pattern of the linear regression was much closer to the observed data than any model realization. During the gap period the RMSE was as high as almost 36 m$^3$/s for the linear regression along the Quevedo River. Nevertheless, as the distance between the stations becomes larger (in other examples or in scattered areas) it is the proposed methodology and its application which might get closer to the typical pattern. For this final comparison, the following linear regression model was used:

$$Q_{Nuevo} = 0.1559 * Q_{Quevedo} - 10.715$$

Eq. III-30

## III.5 Concluding remarks

An alternative technique, for characterizing patterns and its application for estimating missing data in streamflow time series has been presented. This methodology has been adapted to a tropical environment, with only two seasons: wet and dry. The algorithm was based on the Hodrick-Prescott filter in combination with Fourier series. In between, two sub-approaches were also proposed: the block homoskedastic and the heteroskedastic. The former assumed a constant standard deviation, one for each season; whereas the latter assumed the same for the dry period but a variable standard deviation as a function of the sampling trend during the wet period. By means of the algorithm, a random stochastic estimation of a lost signal could be achieved, where the inclusion of seasons introduced some physical knowledge that complemented the mathematical approach.

The pattern characterization was well defined by the HPF. Disregarding typical differences from one year to other, the decomposed signal followed the trend of the original values. Nevertheless, a worth noting setback was experienced when using the Fourier series for projecting the trend (m) where a phase-shift was detected. Further research is required in this regard, finding alternative options that may prevent this displacement or may take in consideration non-cyclical series.

Notwithstanding the difficulty to solve the phase shift impasse, there were still some important remarks to highlight about the performance of the model application. In fact, the methodology appeared not to fail in reproducing the rising tendency and high discharge pattern during the wet season. As for the cases of dry season, the pattern of the estimated series (random realizations) were close enough to the real signal, showing low values and a low-magnitude trend compared with the wet period, as expected. Furthermore, the transition period from the rainy to dry season was successfully represented by the application. The bands captured considerably well the observed data range, of course with the exception of extremes, whose analysis is beyond the scope of the present work.

Finally, for the converse case, the proposed application performed much better. It easily detected the seasonal change and the order of magnitude of the dry period as well as other relevant statistical features.

The sampling period length did not seem to play an important role on the final results when varying from 2 to 3 or 5 or 6 years, for instance. Additionally, better results were obtained when this length was a multiple of 365 (days) or very close to. It is important to recall that this application has been focused on mid-term gaps (e.g. some months). Short-term cases (e.g. few days) can be solved using standard deterministic approaches whereas the solution for long-term periods (e.g. beyond 1 year) may remain impractical.

In opposition to a linear regression between two stations, it was clear that the latter usually works much better than the proposed methodology. Nonetheless, it should be emphasized that especially in developing countries it is not always possible to count on available data in both stations during the same period or to build up a rainfall-runoff model for comparison.

In general, this proposed tool never meant to fully match the original time series, but mainly the pattern and to establish the confidence bands. If longer discharge series were available, the HPF might even improve its performance through a longer learning time and perhaps obtain narrower estimation of bands. In a nearby future there is room for more research that can still be added to this early initiative. Examples might be explored on how the gain is calculated inside the HPF and alternatives for the seasonal separation. A final suggestion may include a non-linear fit (e.g quadratic) for the heteroskedastic variance approach to smooth the transition period between seasons, implying thus a re-estimation of the MLE model.

# IV Ecohydraulics modelling of the AdM wetland-river system[*]

## IV.1 Introduction and justification for modeling.

Some recent approaches highlight the importance of riparian wetlands for their outer river catchment (Bendjoudi et al., 2002; Blackwell et al., 2006; Hattermann et al., 2006). Other works reported embedded models into a catchment framework (Krause and Bronstert 2004; Krause and Bronstert 2005; Hattermann, Krysanova et al. 2006), conceived under the European Water Framework Directive (EC, 2000a). In a similar context, some assessments on water quantity and quality in wetlands within a river basin have been published, mostly in the northern hemisphere (Arheimer and Wittgren 2002; Wang, Shang et al. 2010), even combining modeling with modern remote sensing technologies (van Griensven et al., 2008).

For these dual systems when there is availability of reliable data, physical, hybrid or data-driven models can be useful to either characterize patterns or determine future trends. Unfortunately data are not abundant everywhere. In tropical zones and developing countries, data scarcity is one of the factors that delay the advance of a proper integrated water management. Hence, in those areas there is a necessity to tackle this apparent *trivial* issue before moving to more complex stages. In this regard, a simple method to study streamflow seasonal patterns and estimate gaps in streamflow time series was proposed in Chapter III. Moreover, it must be noted that so far, within and around both the AdM and the Vinces river catchment, very few modeling-based studies have been carried out. Several causes might have contributed to this fact: lack of interest from the governmental entities, few infrastructure works or irrigation programs that may trigger at least temporary data collection campaigns, absence of eco-touristic plans, inaccessibility, etc.

---

[*] Parts of this chapter are based on: Arias-Hidalgo, M., Villa-Cox, G., van Griensven, A., Solórzano, G., Villa-Cox, R., Mynett, A.E., Debels, P. (2012). "A decision framework for wetland management in a river basin context: the "Abras de Mantequilla" case study in the Guayas River Basin, Ecuador", Journal of Environmental Science & Policy, Sp. Ed. (accepted for publication), doi:10.1016/j.envsci.2012.10.009.

With the aim to understand in a better way this wetland-riverine system, a quantitative analysis has been envisaged. However, data were mostly available outside the AdM, specifically in the upstream catchment, i.e. the Vinces-Quevedo and the Daule-Peripa Dam surroundings. Yet again, these data mostly responded to a very particular interest due to the current and future hydropower projects and not to a continuous monitoring campaign from the public sector. Hence, there is a necessity to characterize the system but also to supply information to the region of interest where there is insufficient data.

This data-carrier system of models should respond to the particular conditions of the system (e.g. the Vinces River), or the flow interactions between the Nuevo River and some of the most relevant wetland bodies and branches. Of course models, even under high data availability, do not hold all answers, but may provide a better insight on how the system is working now, how it may respond to future conditions, and most importantly, how the system's responses can be improved via the application of suitable management options.

## IV.2 The modeling framework

In accordance with the DPSIR chains previously defined in chapter II.3, the goal of modeling activities was to assist in the simulation of some processes that take place in and around the wetland. With this in mind, a second stage was to provide information to decision makers prior to policy making and management guidelines (the main objective of the project itself although not of this thesis). The DPSIR indicators for the Ecuadorian case study, from the hydrological point of view, can be summarized in two: water quantity and water quality.

The Daule River as well as the Vinces and the Chojampe Rivers flow southwards (Fig. IV.2-1). Most of the system is still natural with the exception of two major existent hydraulic works: the Daule-Peripa reservoir (5.5 $Km^3$ of capacity) (Arriaga, 1989) which is used for irrigation, flood control in Daule River and deviates 18 $m^3$/s outside Guayas Basin towards La Esperanza Dam, especially during dry season (May to December) The second one is the Chongón project supplying 44 $m^3$/s that goes to the Santa Elena Peninsula, mainly for irrigation and water storage purposes.

The Vinces and the Nuevo River are quite important for the Abras de Mantequilla ecosystem. This is because of an interconnection between Vinces and Nuevo River which allows water from the former to reach southern Abras and viceversa, apparently depending on the seasonality. At the beginning of the rainy season (mid-December, January) water flows from the Nuevo River to the wetland in the Chojampe subbasin. Conversely, when the dry season approaches (normally May) it has been observed in field that AdM drains onto the Nuevo River until equilibrium may be reached thus water stages possibly remain almost constant due to a natural embankment (section IV.5).

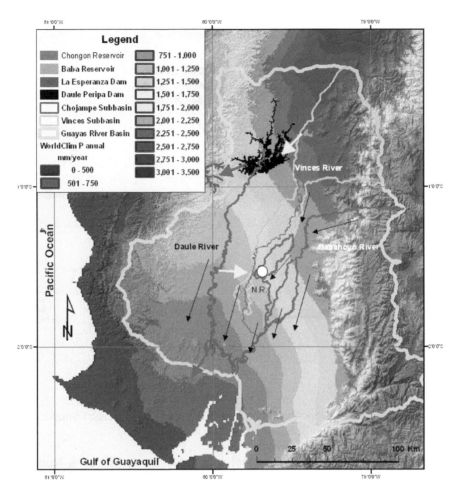

Fig. IV.2-1 Guayas River Basin (Abras de Mantequilla wetland in white), Nuevo River (NR). Current projects (Red arrows). Future projects (yellow arrows).

In order to perform an analysis of indicators and evaluate the management solutions proposed during some stakeholder workshops, some modeling activities were foreseen as shown in Fig. IV.2-2. The modeling domain as well as the connection spots where there is transfer of information is shown in Fig. IV.2-3.

The model chain started with two rainfall-runoff model in the Vinces and Chojampe catchments, with the help of HEC-HMS (Sharffenberg and Fleming, 2010). These models made use of the available rainfall data and computed runoffs at each catchment's outlet.

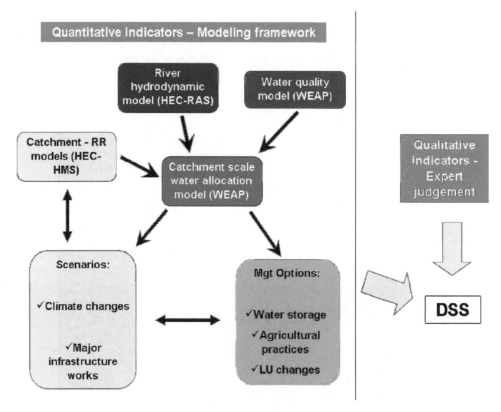

Fig. IV.2-2 Modeling framework and expert judgement towards a Decision Support System.

With that output, a hydrodynamic model was setup along Vinces & Nuevo rivers that included the interactions between the latter and the Abras de Mantequilla system, using HECRAS (Brunner, 2010). The hydrodynamic model was necessary because as mentioned before, there is a time dependant water exchange between the Vinces River and the Abras de Mantequilla wetland through the Nuevo River. The rainfall-runoff and hydrodynamic simulations provided later on the input for a final water allocation model using WEAP (SEI, 2009). For some river stations, such as *Quevedo at Quevedo* and *Nuevo DD Vinces*, streamflow characterization and gap analysis was performed (Chapter III) before being considered as proper boundary condition spots or calibration points for the river routing model.

An assessment on water quality parameters was carried out also in WEAP to compute the water quality index (WQI) in the wetland area. The main WEAP model's target was to undertake a final comparison between the proposed management solutions in the wetland. Simultaneously, the evaluation of the impact of climate changes on the system was performed using rainfall projected time series by PIK (Potsdam Climate Institute, Germany). As a final simulation loop to evaluate the *Business As Usual* (BAU) scenario including climatic variations and infrastructure works, these projected time series were re-incorporated in the aforementioned two HEC-HMS models in Vinces and Chojampe rivers. Finally, those new outputs were considered by the WEAP simulation.

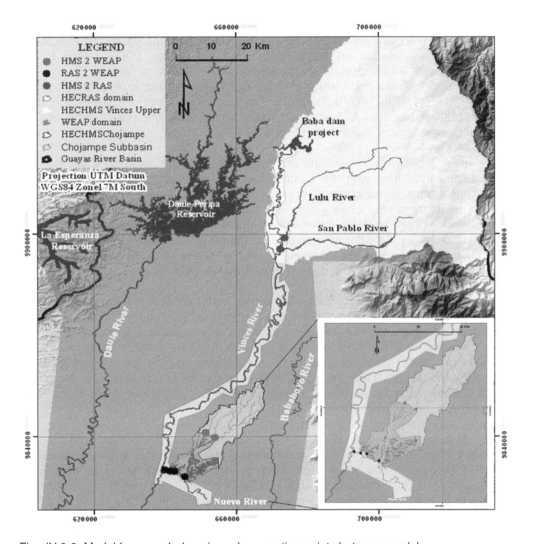

Fig. IV.2-3 Model framework domain and connection points between models.

## IV.3  Rainfall-runoff models: setup and results

### IV.3.1  Vinces upper subcatchment

In order to provide information about lateral inflows to the further hydrodynamic and water allocation models, a rainfall-runoff simulation was required. The Lulu and the San Pablo Rivers are important because they might help mitigating the effects that Baba dam may cause on the Vinces River (Efficacitas, 2006). Due to this assumption, it was important to determine their inflows to Vinces River. The Vinces' upper basin (total area 3416.53 $Km^2$) was divided in 6 subbasins (Fig. IV.3-1). HEC-HMS (Sharffenberg and Fleming, 2010) was the software tool used for the computations. A descriptive flowchart showing the HMS typical internal calculation process is illustrated in Fig. IV.3-2.

The main target was to calculate the hydrographs at the confluence spots of the mentioned tributaries with the Vinces River. It must be remarked that at those tributaries' outlets, the streamflow resolutions were not daily. While at the Lulu River mouth there was a flow gauge in a distant past (up to 1987) with only monthly resolution, at the San Pablo River's mouth there was no record at all. Being thus a justification to simulate the river discharges, the HMS outcomes were used as data for both a hydrodynamic simulation and a water allocation model for the downstream area that included the wetlands inside the AdM area.

The chosen method for accounting the losses from rainfall was the so called Deficit and Constant Loss. This technique is commonly applied due to its simplicity and because it does use few variables to setup a simulation. Besides, it can be conveniently used for long term analyses such as the present case i.e. 1 year (Feldman, 2000). The surface water information required for this rainfall-runoff model are shown in Table IV.3-1. With the help of HEC-GEO-HMS, a GIS-based tool (Fleming and Doan, 2009) it was possible to compute most of those basic inputs.

The constant loss (a.k.a. Phi Index) accounted for the sum of surface storage, canopy interception, infiltration, evapotranspiration, and soil moisture, variables whose measurements were insufficient or not measured at all in this upper catchment. This composed index was estimated for each subcatchment depending on the soil type (Skaags and Khaleel, 1982; USDA, 1986). To determine the soil classification, part of the Curve Number methodology (CN) was employed. A method from the US Natural Resources Conservation Service (NRCS), former US Soil Conservation Service (SCS), it labels soils in four categories, according to their infiltration capacities, based on manifold surveys across the United States' watersheds (USDA, 2004):

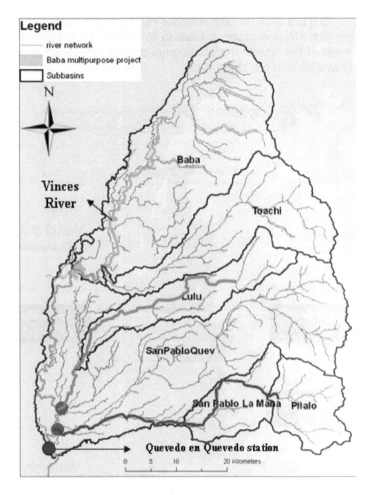

Fig. IV.3-1 HMS schematization for the Vinces upper catchment. Calibration point (blue dot mark) and joint locations with tributaries (purple dots).

- Horizon A: Very high infiltration potential, e.g. gravels and loose sands.

- Horizon B: Moderate infiltration potential, e.g. sandy soils.

- Horizon C: Low infiltration potential (i.e. moderate runoff capacity), clays in general.

- Horizon D: High runoff capacity (expansive clays).

Since the model was aimed to simulate a continuous rainfall regime and not a single event, the Deficit and Constant loss was preferred over the CN method. Although recently the CN variant has been re-adapted for such simulations

(Geetha et al., 2007), it may require a larger number of parameters and surveyed variables in the optimization process leading to higher computational time, beside the fact that most of the typical meteorological data (e.g. evapotranspiration) were not sufficiently available in the study area.

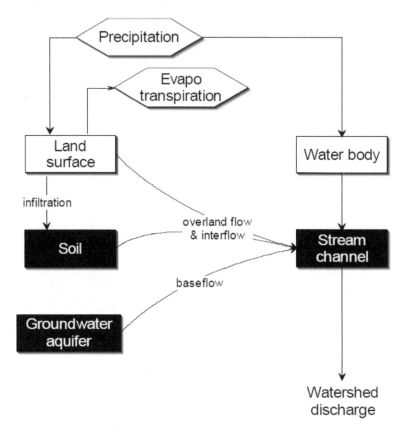

Fig. IV.3-2 Typical HEC-HMS representation of watershed runoff (Feldman, 2000).

Landuse maps (CEDEGE, 2002) were the basis to compute the percentage of imperviousness for each subbasin using the CN variant.    Initial deficits were estimated around 1 mm.  In general the latter have no strong influence on a long term simulation.  As the case study deals with large watersheds, the direct runoff modeling was calculated by means of the lag-time method (USDA, 2004) in spite of other methods such as the kinematic wave (Feldman, 2000).  Nevertheless, since most of the available techniques are empirical or not entirely applicable to all cases, the lag time values were anyway recomputed during the calibration process.

Point rainfall was spatially distributed using cubic splines.  The weights for each subbasin where obtained by merging the areas of influence with the drainage areas previously delineated using HEC-GEOHMS.    The contribution of groundwater flow was estimated using the recession approach. Table IV.3-3 shows the involved variables. One of them was the recession constant, defined as $Q_{bi+1}/Q_{bi}$ in the typical baseflow exponential model. The second was the "ratio to peak" $Q_b/Q_p$ where $Q_p$ is the discharge peak. When any ratio is smaller than this threshold, the total discharge is mainly driven by groundwater flow. The third was the initial discharge that, as the former two, could be derived from historical discharge hydrographs.

Table IV.3-1 Surface water variables for Vinces River model in HEC-HMS

| Subbasin | Area (km$^2$) | Constant loss rate (mm h$^{-1}$) | % Imperviousness | Lag time (min) |
|---|---|---|---|---|
| Baba | 925.2 | 3.6 | 4.1 | 1700 |
| Toachi | 504.8 | 3.1 | 4.5 | 1407 |
| San Pablo - Quevedo | 1290.3 | 4.0 | 5.8 | 1507 |
| Pilaló | 212.9 | 2.8 | 4.7 | 1186 |
| San Pablo - La Maná | 190.0 | 3.3 | 3.8 | 1507 |
| Lulu | 293.4 | 4.3 | 5.3 | 1326 |

Table IV.3-2 Gage weights for San Pablo-Quevedo subbasin.

| San Pablo - Quevedo Subbasin | |
|---|---|
| Gage Name | Weight |
| Inmoriec Vergel | 0.31 |
| Pichilingue | 0.05 |
| Pilaló | 0.22 |
| Puerto ila | 0.00 |
| San Antonio Delta Pate | 0.30 |
| San Juan La Maná | 0.10 |
| Union 71 | 0.02 |

The time control parameters for computation were taken from January 1$^{st}$ to December 31$^{st}$ 2006. Important to note is that the time step was daily because there was not finer temporal resolution available for rainfall and river discharge observations, needed for calibration purposes. As for the calibration, the most sensitive variables were, as expected, the constant loss index, the lag time, the baseflow threshold ratio and the recession constant. The Univariate Gradient was used for optimizing the calibration process, with a tolerance of 1% or maximum 100 iterations. Finally, the Sum Squared Residuals was the chosen objective function, still frequently employed in this respect (Diskin and Simon, 1977):

Table IV.3-3 Baseflow parameters for HEC-HMS

| Subbasin | Initial Discharge (m³/s) | Recession constant | Ratio to Peak |
|---|---|---|---|
| Baba | 25.5 | 0.79 | 0.78 |
| Toachi | 1.8 | 0.79 | 0.76 |
| San Pablo - Quevedo | 6.3 | 0.85 | 0.67 |
| Pilaló | 1.0 | 0.95 | 0.85 |
| San Pablo - La Maná | 4.3 | 0.76 | 0.65 |
| Lulu | 2.6 | 0.93 | 0.70 |

$$Z = \sum_{i=1}^{N} \left[ Q_{obs}(i) - Q_{sim}(i) \right]^2$$

Eq. IV-1

where Z is the objective function to be minimized; N is the total number of compared instances; Qobs is the observed streamflow and Qsim is the simulated one. In order to check the efficiency of the model, the Nash-Sutcliffe coefficient was computed for each location in which reliable observations were available on a daily time step t (Eq. IV-2 and Table IV.3-4):

$$E = 1 - \frac{\sum_{t=1}^{N} (Q_{obs}^t - Q_{sim}^t)^2}{\sum_{t=1}^{N} (Q_{obs}^t - \overline{Q_{obs}})^2}$$

Eq. IV-2

where:
E is the Nash-Sutcliffe number; and,

$\overline{Q_{obs}}$ is the average of the observed discharge values.

Table IV.3-4 Nash-Sutcliffe values for some subbasins within the Vinces HMS model.

| Subbasin | Nash-Sutcliffe Coefficient |
|---|---|
| Toachi | 0.69 |
| Baba | 0.67 |
| Pilaló | 0.84 |
| SanPablo - Quevedo | 0.81 |

The Fig. IV.3-3 shows an example of hydrograph comparison between observed and computed values, at the *Quevedo at Quevedo* streamflow station. It was observed that although some of the observed peaks were not accurately matched by the simulation, at least the trend and some other peaks are very well represented. HEC-HMS computed several flow peaks in this period responding to the respective precipitation events, for instance in the Baba River subbasin (peaks in May, October and November 2005, Fig. IV.3-4), one of the tributaries of the Quevedo-Vinces river. It can be observed that the model responded accordingly to every rainfall event. Although the period from May to December has been labeled as "dry", there are however two particular peaks in June and November, consequence of some isolated rainfall events in the Andes' foothills (Fig. IV.3-3). According to local experiences, there might have been some reliability problems about the discharge measurements provided by the Institute of Hydrology in Ecuador (INAMHI), particularly during the dry season. Fortunately, this has not been the case for most of the rainfall measurements. Despite these *suspicious* mismatches, the Nash-Sutcliffe number was acceptable taking in consideration the simplicity of the model.

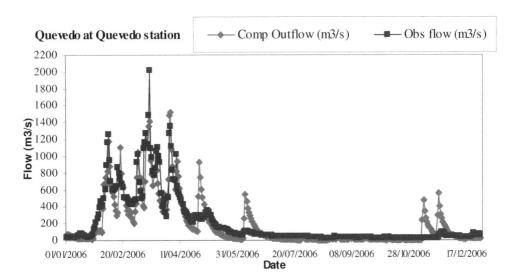

Fig. IV.3-3 Hydrograph comparison for the Vinces HMS model.

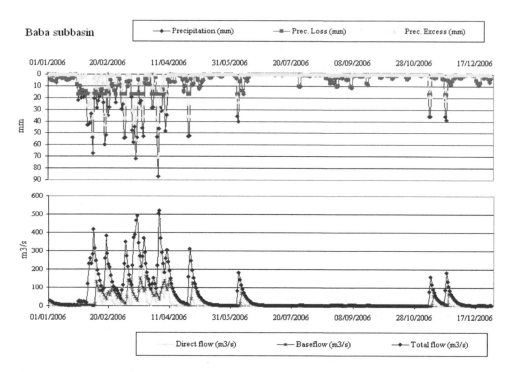

Fig. IV.3-4  Rainfall hyetograph and flow hydrograph at Baba subbasin.

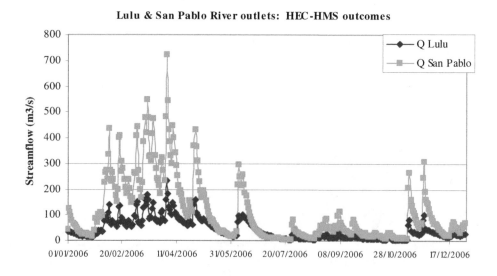

Fig. IV.3-5  River discharges at Lulu and San Pablo outlets, from HEC-HMS.

Finally, the discharges coming from the Lulu & San Pablo River were part of the boundary conditions to be used for the river routing model (described shortly). It was observed that the San Pablo contributes 2.3 times more than the Lulu to the Vinces River (Fig. IV.3-5), keeping this pattern along the whole year, mainly due to the larger size of its watershed.

## IV.3.2  The Chojampe-AdM subcatchment

For the rainfall-runoff model in the Chojampe subbasin (around 290 Km$^2$), eight micro-basins were considered. The subbasin delimitation and main features are shown in Fig. IV.3-6. The biggest water bodies in the wetland area are located in the three southernmost catchments, namely, *El Recuerdo*, *Abanico* and *Abras de Mantequilla*. In addition, an important spot was the *Chojampe 2/Agua Fria* outlet, since that represented one of the link points between the HMS and the WEAP model (Chojampe river's headflows).

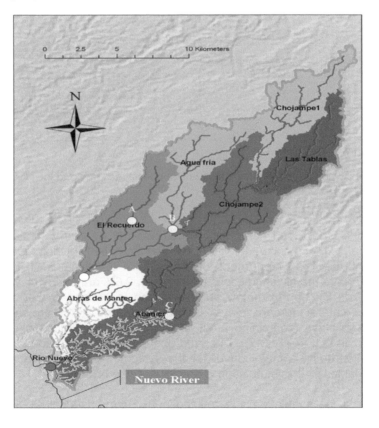

Fig. IV.3-6  Schematization of HMS model for the Chojampe subbasin. Transfer points to WEAP (yellow dots). Calibration point (white dot).

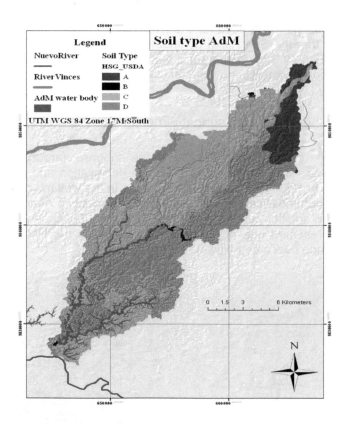

Fig. IV.3-7 Soil type variability in the Chojampe subbasin, according to the USDA classification (USDA, 2004).

For this model, the runoff computation methodology was very similar as for the Vinces model. The spatial variability of the soil types in the Chojampe Basin took into account the US Department of Agriculture classification (USDA, 2004). Around 12.8% are mainly sandy soils (Horizon A), 2.6% are sandy clays and loamy sands (Horizon B); approximately, 48.2% belong to horizon C, i.e. clays, loamy clays, and in general inorganic clays. Ultimately, around 36.4% are saturated soils and expansive clays (containing most likely illite and montmorillonite), classified thus as horizon D (Fig. IV.3-7). The land use distribution was already discussed, as degradation being a major driver for the system. A summary of the different involved variables for this simulation is shown in Table IV.3-5.

Like in the Vinces simulation, the training period comprised all 2006. The only available calibration point was at *El Recuerdo* village, located at the namesake microbasin outlet point. As a result of the minimization of the error function, the Nash-Sutcliffe number was 0.74 (Fig. IV.3-8) which in general terms would means a good performance. Important to recall is that the yellow dots in Fig. IV.3-6 indicated the interest locations where the discharge was calculated (A, B, C). These points were selected because there the river influence on the wetland is already very small. Therefore, the rainfall-runoff analysis focused only in the upstream part of those locations, as they served as boundary conditions for a further water allocation model in the wetland-Nuevo River area. Ultimately, a plot-summary of these resultant streamflow series is depicted in Fig. IV.3-9. As expected, the highest volumes were computed at B, due to the larger number of drained micro-basins.

Table IV.3-5 Surface & groundwater features for the Chojampe-AdM model

| Micro-basin | Area (km²) | Initial Abstr. (mm) | Constant loss rate (mm/h) | Imperv. (%) | Lag Time (min) | Initial Flow (m³/s) | Recession Constant | Ratio to Peak |
|---|---|---|---|---|---|---|---|---|
| Las Tablas | 27 | 1.14 | 4.08 | 5 | 302.4 | 0.1 | 0.92 | 0.65 |
| Chojampe 1 | 29 | 0.95 | 3.7 | 6 | 326.4 | 0.1 | 0.94 | 0.67 |
| Chojampe 2 | 39 | 0.53 | 1.61 | 8 | 456.0 | 0.1 | 0.92 | 0.65 |
| Agua fría | 31 | 0.56 | 2.4 | 9 | 323.4 | 0.1 | 0.92 | 0.65 |
| El Recuerdo | 41 | 0.53 | 2.04 | 9 | 357.6 | 0.1 | 0.91 | 0.65 |
| Abras de Mantequilla | 24 | 0.48 | 1.42 | 12 | 265.8 | 0.1 | 0.86 | 0.67 |
| Abanico | 49 | 0.48 | 1.31 | 11 | 230.4 | 0.1 | 0.94 | 0.67 |
| Río Nuevo | 1 | 0.53 | 0.78 | 30 | 67.8 | 0.1 | 0.84 | 0.67 |

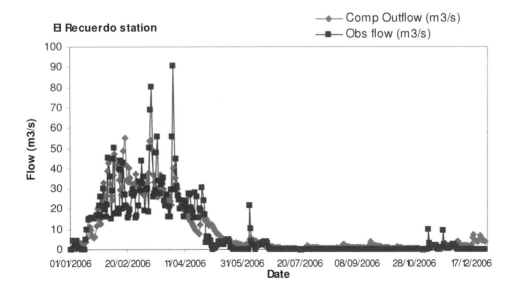

Fig. IV.3-8 Hydrograph comparison, Chojampe HMS model.

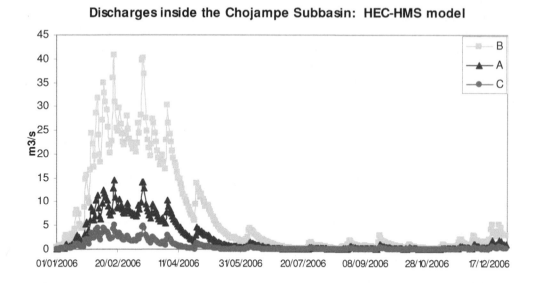

Fig. IV.3-9 Flows at the transfer points between HEC-HMS and WEAP. Chojampe model.

## IV.4 The use of TRMM data for the Vinces upper catchment*

### IV.4.1 Current status of the research and worldwide experiences

Nowadays, remote sensing is immensely useful to improve our understanding of spatio-temporal variation of rainfall, particularly for data scarce regions. In this regard, the Tropical Rainfall Measuring Mission (TRMM) (Kummerow et al., 1998; Simpson et al., 1988), an initiative of the US Space Agency (NASA) and the Japanese Aerospace Exploration Agency (JAXA), is instrumental in shaping the research related to the use of satellite based rainfall products in hydrological studies (http://trmm.gsfc.nasa.gov/). TRMM is operational since November 1997 and is releasing products since 1998.

As its name indicates, the TRMM mission covers only the tropical zone, i.e. between the latitudes 50°N and 50°S. The current spatial resolution is 0.25°. The satellite possesses five instruments on board (i) the Precipitation Radar (PR) which

* Chapter section based on: Arias-Hidalgo, M., Bhattacharya, B., Mynett, A. E., van Griensven, A. (2012), "Experiences in using the TRMM data to complement raingauge data in the Ecuadorian coastal foothills", Journal of Hydrology and Earth System Sciences, Discuss., 9, 12435-12461, doi:10.5194/hessd-9-12435-2012.

records the intensity, distribution, type of the rain, the storm depth and the snowmelt height, with a swath width of 215 km; (ii) the Microwave Imager (TMI) which senses the microwave energy emitted by the planet and the atmosphere, with a width of 760 km; (iii) the Visible and Infrared Scanner (VIRS) which measures the radiation originated in the planet in several spectral zones, with a swath width of 720 km wide; (iv) the Clouds and the Earth's Radiant Energy System (CERES) which measures the energy levels in the highest region of the atmosphere as well as on the Earth's surface; and (v) the Lightning Imaging Sensor (LIS), the intra-cloud and cloud-to-ground lightning detector.

A large number of publications have reported worldwide experiences on the use of TRMM products (Nicholson 2005; Hughes 2006; Collischonn, Collischonn et al. 2008; Wong and Chiu 2008; Buarque, de Paiva et al. 2011; Rollenbeck and Bendix 2011), particularly the 3B42 type (Huffman et al., 2007). In this regard, two lines of research are noticed. The first one have been focusing on comparing the TRMM rainfall data with the raingauge data, either to study the spatial and temporal variability, or to test the validity of the TRMM products. The second line of research has investigated the potential use of the TRMM rainfall data as an independent data source or in complementing raingauge data for hydrological studies.

There are important works related to the first category. For instance, in the arid environments of southern Africa, Nicholson (2005) and Hughes (2006) reported that the TRMM data overestimated the raingauge data in every comparison based on a monthly scale. Other interesting cases are the ones reported by Bell & Kundu (2003). They also compared on a monthly basis and recognized that even in densely gauged networks there were large differences between ground data and TRMM data. A number of studies have concluded that a comparison on annual time scale yields very good results but with finer temporal scales the error starts increasing. At daily or weekly time resolution, error values around 50% have been reported (Huffman et al., 2010; Olson et al., 1996; Wilheit, 1988). Around the globe other examples of comparisons have been carried out, in places such as Hong Kong (Wong and Chiu, 2008), the Brazilian part of the Amazon River Basin (Buarque et al., 2011), Indonesia (Vernimmen et al., 2012) and countries with poor data conditions such as Ghana (Endreny and Imbeah, 2009).

The location of the selected study area seems to strongly influence the comparison performance. Publications whose case studies deal with oceanic environments or flat areas (e.g. Amazon Basin) have reported very good matches between the data from raingauges mounted on buoys and the TRMM data (Adler et al., 2000; Bowman, 2005). In studies about locations with higher altitudes and particularly in the foothills of mountainous regions (e.g the Andes), there were notorious differences between the two sources of data (Tian and Peters-Lidard, 2010). In this regard, under the orographic effect TRMM might show lower values than the gauge rainfall (Dinku et al., 2010). To worsen the scenario, these areas are frequently the most unattended by the national weather agencies in terms of data availability and reliability. Given this background, there is an extreme heterogeneity and uncertainty of the spatio/temporal distribution of the convective rainfall (Bendix et al., 2009). For this challenge, TRMM and in general satellite data may contribute for a better comprehension of the spatial and temporal pattern features of precipitation, in particular if space borne and gauge data complement to each other (Rollenbeck and Bendix, 2011). This possibility still needs to be investigated in areas with large spatial variability of rainfall.

A second group of researchers have gone beyond data comparisons. They have used the satellite products as a new input for rainfall-runoff models and then compared the simulation results with that of the original model. Noteworthy examples are the models developed in California (Guetter et al., 1996; Yilmaz et al., 2005) where flow simulation and soil water estimates were undertaken at a meso-scale basin using the GOES (Geostationary Operational Environmental Satellite) data. Although the simulation outcomes, when compared with the conventional hydrological simulation, were not very accurate, the authors were able to demonstrate a procedure of combining manifold data sources. Possible sources of error may have been (i) the quality level of the GOES atmospheric correction algorithms at that time; and (ii) the fact that the precipitation estimations (Yilmaz et al., 2005) were aimed to ungauged basins, hence involving a large uncertainty. In the Tapajós River (Amazon Basin, Brazil) spatial distribution of rainfall as well as daily comparisons of different data sources for hydrological simulations have been investigated. Firstly using only raingauge observations; and secondly, integrating these point measurements with TRMM (Collischonn et al., 2008). These comparisons gave support for large-scale rainfall-runoff and further hydrodynamic simulations (Paiva et al., 2011).

The literature suggests the promising possibility of complementing the rainfall data from raingauges with that from TRMM in hydrological studies of data scarce regions such as the Vinces Basin in Guayas River Basin in Ecuador. This paper presents a simple procedure to combine the two aforementioned sources in hydrological simulation of the existing rainfall-runoff model of Vinces Basin. (Section IV.3.1).

## IV.4.2 Methodology and Results

Among several TRMM data products, the 3B42 were used in this study as it has been recommended by previous researchers (Winsemius 2009; Dinku, Connor et al. 2010; Almazroui 2011; Vernimmen, Hooijer et al. 2012). At first, the 3B42 data were downloaded from the geo-data website of the King's College in London (http://geodata.policysupport.org/rainfall-timeseries). The three-hourly data throughout a time span of 8 years (1999-2006) was available, which allowed the aggregation of the values to a daily, monthly and yearly resolution.

Annual rainfall from rain ground stations and satellite, averaged over the available period 1999-2006, were computed at their respective measurement points. Adopting the Inverse Distance Weighting (IDW) for interpolation, an average spatial distribution of annual rainfall is shown in Fig. IV.4-1a & b. The ground-based map indicated an increasing pattern principally towards the north. Such a trend was somewhat also captured by the TRMM-based map, although its order of magnitude was 50-65% smaller than the raingauge representation. It is noted that upper-right part of the catchment shows the lowest density of ground based measuring stations. This fact corroborates the concerns about possible high uncertainties that may be associated with rainfall estimation across foothill areas (Paiva et al., 2011).

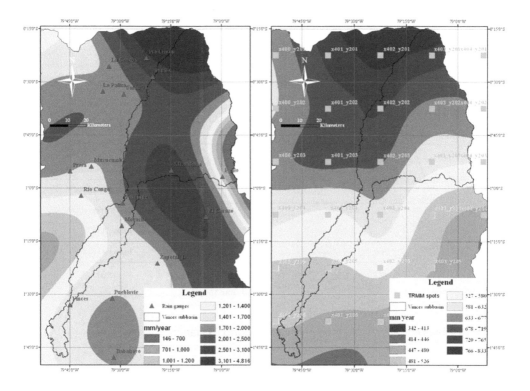

Fig. IV.4-1 Annual average rain field comparison between: a) ground stations (left); and b) TRMM data (right), during 1999-2006.

Several time scales were considered for bias correction: annual, seasonal, monthly, etc. The monthly resolution proved to be the finest one with still a high correlation between the two rainfall data sources (to be described shortly). Monthly bias correction has been adopted in previous researches (Bell and Kundu 2003; Hughes 2006; Rollenbeck and Bendix 2011; Vernimmen, Hooijer et al. 2012). Beyond that resolution, at the rain stations and at daily scale it was common to find poor correlations between the raingauge observations and the raw TRMM data ($R^2$ < 0.30).

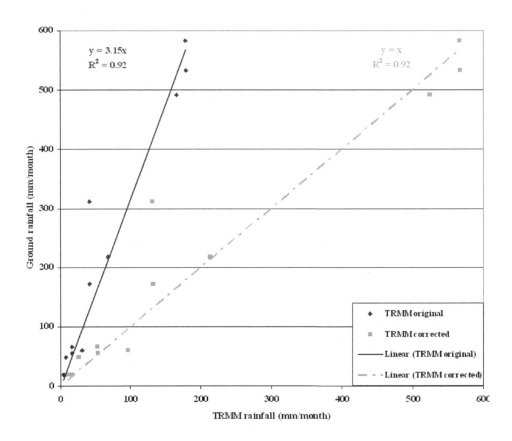

Fig. IV.4-2 Bias correction at a monthly scale.  Puerto ila station.

Frequently the raingauge emplacements and the TRMM grid cell centres do not coincide.  As a consequence, the average monthly TRMM data (at the grid cells) had to be estimated at the raingauge locations (IDW was used once again).  Thus, the average monthly rainfall values for the study period (1999-2006) measured at each rain gauge location were compared against their TRMM interpolated counterparts.  The following equation expresses a relationship between the raingauge and the *uncorrected* TRMM monthly values, at a certain location *i*:

$$TP_{i,m} = K * \mathrm{TRMM}_{i,m}$$

Eq.  IV-3

where:
*K* is the bias factor at the raingauge location *i*;

$TRMM_{i,m}$ is the uncorrected monthly rainfall (mm/month), obtained from the satellite data and estimated at the raingauge location $i$ during the month $m$; and,
$TP_{i,m}$ is the total rainfall at raingauge $i$ during the month $m$ (mm/month), from ground observations.

An example of this correlation can be seen for the *Puerto ila* station (Fig. IV.4-2). Table IV.4-1 shows the expanded results of this annual comparison (based on monthly scale adjustment). In general, it was observed a high correlation at monthly scale ($R^2$ = 0.81 in average). In that regard, Fig. IV.4-3 illustrates a graphical comparison between the ground observations, the uncorrected and corrected TRMM data. In order to assess the validity of the bias correction, the relative bias and the Root Mean Square Error (RMSE) were calculated as follows:

Table IV.4-1 Bias correction, TRMM vs. ground data, annual rainfall based on monthly correction.

| Validating station | Ground data, annual rainfall (mm yr$^{-1}$) | Uncorrected TRMM | | | | | Corrected TRMM | | |
|---|---|---|---|---|---|---|---|---|---|
| | | Annual Rainfall (mm yr$^{-1}$) | Rel. bias (%) | RMSE (mm yr$^{-1}$) | Monthly bias corrector | $R^2$ | Annual Rainfall (mm yr$^{-1}$) | Rel. bias (%) | RMSE (mm yr$^{-1}$) |
| Puerto ila | 2578.1 | 757.6 | 70.6 | 206.4 | 3.15 | 0.92 | 2389.1 | 7.3 | 56.7 |
| San Juan La Maná | 2805.8 | 600.0 | 78.6 | 256.0 | 4.28 | 0.89 | 2571.0 | 8.4 | 76.4 |
| Pichilingue | 1858.3 | 595.0 | 68.0 | 149.6 | 4.29 | 0.85 | 1654.5 | 11.0 | 63.2 |
| Murucumba | 1738.9 | 693.0 | 60.1 | 125.3 | 2.18 | 0.87 | 1508.3 | 13.3 | 57.3 |
| Pilaló | 1095.1 | 565.5 | 48.4 | 60.3 | 1.85 | 0.73 | 1045.5 | 4.5 | 34.5 |
| Chiriboga | 4653.7 | 759.3 | 83.7 | 356.7 | 5.66 | 0.63 | 4295.6 | 7.7 | 108.3 |
| Puerto Limón | 2527.5 | 839.9 | 66.8 | 184.4 | 2.56 | 0.78 | 2151.3 | 14.9 | 86.8 |
| Unión 71 | 1983.6 | 769.1 | 61.2 | 141.1 | 2.25 | 0.82 | 1728.0 | 12.9 | 69.2 |
| La Cancha | 1730.8 | 678.3 | 60.8 | 125.6 | 2.18 | 0.75 | 1481.5 | 14.4 | 71.9 |
| La Palizada | 1805.8 | 577.6 | 68.0 | 150.6 | 2.87 | 0.87 | 1657.0 | 8.2 | 58.2 |
| El Corazón | 2391.1 | 509.2 | 78.7 | 217.5 | 4.47 | 0.83 | 2274.1 | 4.9 | 77.5 |

$$relative\ bias\ (\%) = \frac{P_{groudst} - P_{TRMM}}{P_{groudst}} * 100$$

Eq. IV-4

$$RMSE(mm / year) = \sqrt{\frac{\sum_{i=1}^{12}(TP_{i,m} - p_{TRMM_i})^2}{12}}$$

Eq. IV-5

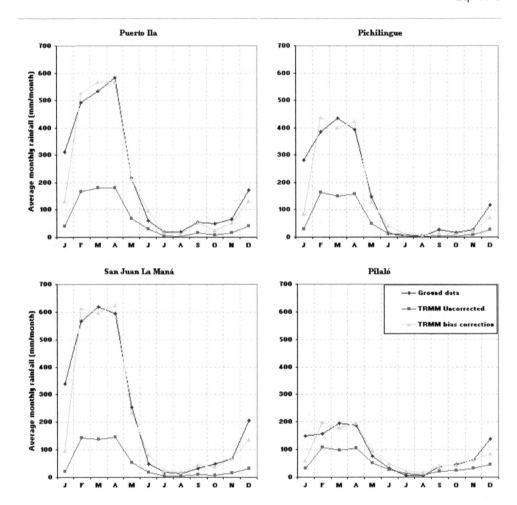

Fig. IV.4-3 Average monthly bias corrected TRMM data over 1999-2006 compared with rain gauges and uncorrected TRMM data..

where:

$P_{Groundst}$ is the annual rainfall from ground observations (mm/year);

$P_{TRMM}$ is the uncorrected and corrected annual rainfall, derived from the satellite data (mm/year);

$p_{TRMM\ i}$ is the monthly rainfall for month $m$, at raingauge location $i$, for both uncorrected and corrected TRMM information (mm/month).

As a further step, the bias adjustment coefficients (K in Eq. IV-3) were spatially distributed across the Vinces upper catchment resulting in a distributed map of correctors (Fig. IV.4-4). As before, the approach was the inverse distance weight, based on the correctors estimated at each raingauge location. As it could have been expected from the differences in annual averages, bias correctors between 2.7 and 3.2 constituted a representative interval for most of the catchment domain, only with the exceptions of those ground stations situated in the uppermost portions of the catchment (close to the water divide). The correspondent bias correction coefficients were estimated for each of the TRMM grid centres and thus the correction took place using the following expression:

$$\mathrm{TRMM}_{\mathrm{corr,\ j,m}} = K'*\mathrm{TRMM}_{\mathrm{j,m}}$$

Eq. IV-6

where:
$K'$ is the monthly bias factor, estimated at the TRMM grid centre $j$;
$TRMM_{j,m}$ is the uncorrected TRMM monthly rainfall at the grid centre $j$ during month $m$ (mm/month); and,
$TRMM_{corr,j,m}$ is the corrected TRMM monthly rainfall at the grid centre $j$ during month $m$ (mm/month).

Compared to the original setup of the rainfall-runoff model (section IV.3), five TRMM-based rainfall stations were thus incorporated to the simulation area (circles in Fig. IV.4-4), two in the lower area and three in the highlands. Because the rainfall-runoff model was built using a daily time step, the satellite corrected monthly values required to be disaggregated to a daily resolution for each new input spot. To achieve this, empirical factors ($f_i$) were derived from the raingauge time series in this way:

$$f_{i,d} = \frac{P_{i,d,m}}{TP_{i,m}}$$

Eq. IV-7

where:
$f_{i,d}$ is the temporal disaggregation coefficient, at the raingauge $i$, for the day $d$ of month $m$;
$P_{i,d,m}$ is the total rainfall at raingauge location $i$ in the day $d$ of month $m$ (from ground observations, mm/day); and,
$TP_{i,m}$ is the total rainfall at raingauge location $i$ during the month $m$, from ground observations (mm/month) as explained in Eq. IV-3.

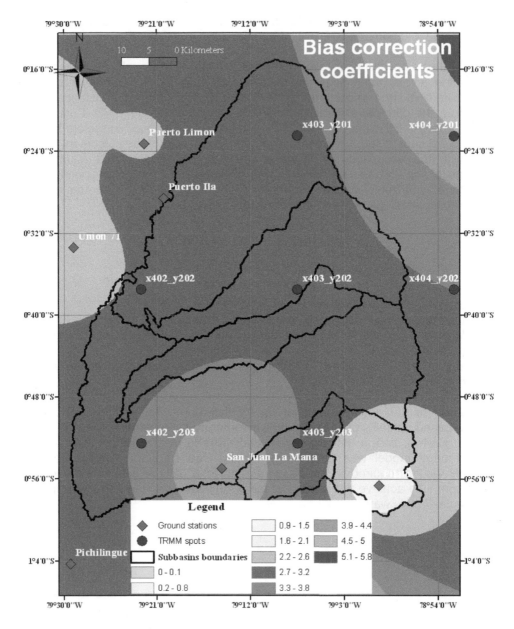

Fig. IV.4-4 Spatial distribution of bias corrector coefficients at monthly scale. Vinces upper catchment..

The $f_{i,d}$ ratios were then applied back to the corrected TRMM monthly values to estimate the daily series (day x, month X) at the satellite grid centres. There, the procedure took the factors from the nearest ground location. The final expression is as follows:

$$TRMM_{corr,j,d} = f_{i,d} * \text{TRMM}_{corr,j,m}$$

Eq. IV-8

where:

$TRMM_{corr,j,d}$ is the disaggregated, daily corrected TRMM monthly rainfall at grid centre $j$ (mm/day).

The iterative process continued until the simulation time span was completed. Finally, in order to illustrate the validity of this simple procedure, an example was taken from the location of the Puerto ila gauge station, as it is shown in Fig. IV.4-5. At daily scale, the correlation at this spot was high enough ($R^2$ = 0.88) given the empirical approach and the large initial bias.

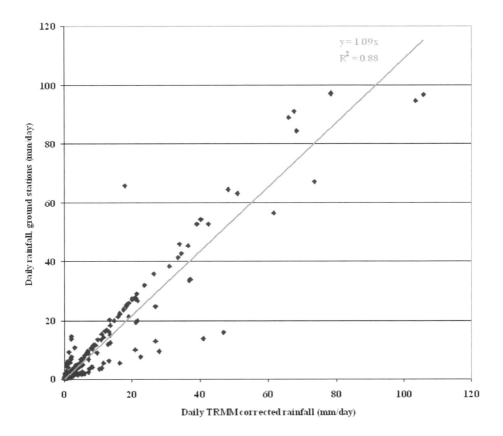

Fig. IV.4-5 Comparison of rainfall data from gauges and TRMM corrected data, at daily scale. Puerto ila station.

## IV.4.3  Performance of complementary TRMM data for the HMS model

In order to assess the usefulness of combining rainfall data from raingauges and TRMM as an alternative data for rainfall-runoff modeling, a new hydrological simulation was executed for the Vinces upper catchment. The enlarged rainfall input scheme is according to Fig. IV.4-4. While the rainfall ground stations daily data series remain unaltered, the new TRMM spots made use of the previously obtained synthetic daily series, corrected from the original satellite data.

This new scheme entailed the recalculation of the average rainfall per subbasin and areas of influence for each station. Some worthy comparisons were then achieved (Fig. IV.4-6). During some peaks throughout the rainy season of 2006 (e.g. 8 February, 5 March, and others), the newly fed model showed higher streamflow values compared to the ground-based data model. Given the previous underestimation of the original model with respect to the observations, this might have implied a sort of improvement on the model performance (8-13%) for the general pattern and some peaks (e.g. the peak of 16 March). However, for other peaks the new model caused a larger error of around 18%. The Nash-Sutcliffe coefficient for the wet period slightly declined, from 0.83 to 0.81.

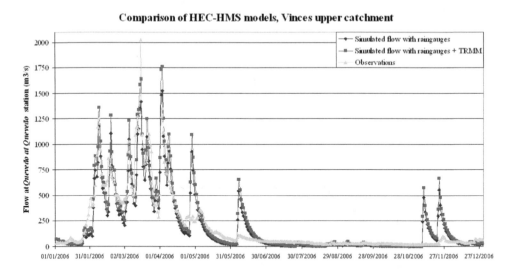

Fig. IV.4-6 Rainfall-runoff simulations with different sorts of precipitation data, Vinces River upper catchment.

Secondly, for the dry season and with the exception of the aforementioned events of May, June and November, the NSC somewhat increased from 0.98 to 0.99. When those events were included, the new simulation showed no improvement compared to the original model due to the overestimation on the original values with respect to the observations. Finally, as an overall yearly view, the Nash Sutcliffe coefficient slightly decreased from 0.81 to 0.76.


## IV.5 The unsteady river flow model

### IV.5.1 Justification

A river routing model was conceived along the Vinces and the Nuevo Rivers, as well as the Abras de Mantequilla main water body. Its main objective was to serve as a "data" carrier, taking information from the upstream part of the basin (Quevedo-Vinces system) to the Nuevo River and then to the wetlands connection point. HEC-RAS, a tool developed by the US Army Corp of Engineers (Brunner, 2010), carried out an unsteady computation in 1D (x direction) using the Saint-Venant equations. Two reasons justify this kind of simulation:

• There is a noticeable interaction between the river and the wetland that depends on the seasonality and needed to be quantified. In general, a hydrodynamic analysis is a fundamental step for a good characterization of any environmental assessment within the wetland as remarked by some previous experiences (Holland, Whigham et al. 1990; Lowrance, Altier et al. 1995; Whigham 1999).

• Steady analyses are not consistent with water retained in floodplains. Since the wetland was involved in the river analysis (storage area), the best option is the unsteady simulation (Brunner, 2010).


### IV.5.2 Model setup and results

Existent topographic data were pre-processed in ArcGIS to produce a Digital Elevation Model (DEM) (Chapter II). Several sources were employed depending on the data availability. For the upstream first 70 Km, the DEM included a recent bathymetry obtained from feasibility studies of the Baba project (Efficacitas, 2006). The following 50 Km were derived from the Shuttle Radar Topographic Mission (SRTM) data (available at http://srtm.csi.cgiar.org/). Finally, for the lower part and the wetland area, a raster file processed by the Ecuadorian Army Geographic Institute (IGM) in scale 1:10000 was incorporated to the overall elevation model (spatial resolution 5m). In total, 174 km & 25 km along the Vinces and Nuevo Rivers, resp.

Cross sections were generated by means of HEC-GEORAS (Ackerman, 2009) using the aforementioned DEM. The space step was fixed in 200 meters to reduce computational costs. Moreover, roughness (Manning) coefficients were assigned to every cross-section, in both main channel and floodplains, according to the existent landuse (e.g. vegetation/crops) and literature (Chow, 1959). Typically, values range from 0.03 to 0.04 for the river reaches and a maximum of 0.06 for floodplains, especially along the Nuevo River (dense vegetation). Because of the absence of abrupt variations along the main channel geometry, the expansion and contraction effects of the Saint-Venant equations were negligible for this unsteady analysis (Brunner, 2010). The total time span was 12 months (January 2nd, December 30th, 2006). Although the only available discharge measurements were daily, the computational time step was fixed in 3 minutes (180 seconds). Because the selected cross-section separation was established in 200 meters, this ensured a Courant number <1 and thus numerical stability along the domain.

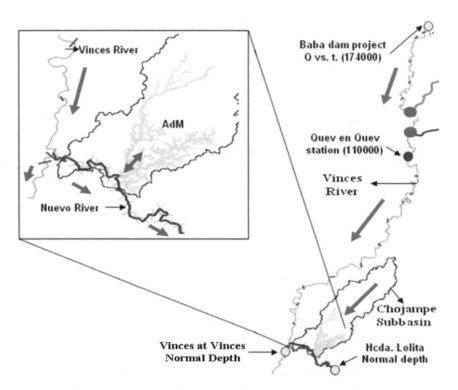

Fig. IV.5-1 HECRAS model geometric schematization. Boundary conditions (yellow dots), calibration point (blue dot). Inflows from rivers: Lulu (orange) and San Pablo (blue).

In order to complete the resultant system of equations from the Preissmann numerical scheme (used by HECRAS), some boundary conditions (BC) were set. Firstly, as upstream BC, a flow hydrograph on the location of the future Baba Multipurpose project (upstream BC for Vinces River, Km 174). Secondly, based on the outcomes from the first rainfall-runoff model, the flow hydrographs from the Lulu & San Pablo Rivers, main tributaries of the Vinces River, were added as lateral inflows (Fig. IV.5-1). Finally, *normal depths*, resulting from given values of friction slopes in the Manning equation, were assigned as downstream boundary conditions in both Vinces Nuevo Rivers (Vinces town and Hcda. Lolita, respectively).

As mentioned before, the most relevant wetlands were simulated as storage areas. Therefore, initial water depths were assigned to each of the three largest water bodies along the left bank of the Nuevo River. These were labeled: West Abras, Central Abras and Main Abras (Fig. IV.5-2). The confluences where these "abras" meet the Nuevo River (they have not been intervened yet) are also the connection spots between the hydrodynamic model and the water allocation model (Fig. IV.2-3), to be described shortly. Based on existing topography, the stage-volume curves were obtained for each wetland (Table IV.5-1). The flow interaction between the river and each body was approximated using simulated trapezoidal weirs whose data can be seen in Table IV.5-2. These "lateral structures" used the Kindsvater-Carter rectangular weir (standard) equation (Kindsvater and Carter, 1959) to compute the inflows and outflows over these natural levees. The expression in mention is as follows:

$$Q = C_e \frac{2}{3} \sqrt{2g}\,(b + K_b)(h + K_h)^{1.5}$$

Eq. IV-9

where:

Q = Discharge (m$^3$/s);
$C_e$ = Dimensionless discharge coefficient;
g = Acceleration of gravity (m/s$^2$);
b = weir width (m);
h = water depth (m);

$K_b$ and $K_h$ are the viscosity and surface tension parameters (m); however, compared with the order of magnitude of b and h, they are very small and hence they were neglected.

The calibration process was mainly focused on the roughness coefficient along the Vinces and Nuevo Rivers. Its time span lasted similarly as in the HEC-HMS model, given the observed discharges at the *Quevedo at Quevedo* station (NSC = 0.81). That comparison can be seen in Fig. IV.5-3. Again the model works well enough during the wet season, with the exception of isolated flow spikes in late April, early June and November, most likely caused by likewise events coming from the Lulu & San Pablo Rivers (boundary conditions). Unfortunately, the field observations failed to register such streamflow fluctuations.

Fig. IV.5-2 Wetland water bodies along the Nuevo River. Example of natural weir (red rectangle).

Table IV.5-1 Stage-volume curves for the Abras wetlands

| West Abras | | Central Abras | | Main Abras | |
|---|---|---|---|---|---|
| Stage (m) | Volume (m3) | Stage (m) | Volume (m3) | Stage (m) | Volume (m3) |
| 10.500 | 0 | 10.000 | 0 | 5.970 | 0 |
| 10.790 | 58 | 10.320 | 200358 | 6.770 | 36280 |
| 11.140 | 131 | 10.790 | 497996 | 7.740 | 364870 |
| 11.560 | 227 | 11.090 | 689979 | 8.900 | 4193935 |
| 12.070 | 367 | 11.900 | 1212514 | 9.624 | 8156494 |
| 12.670 | 55651 | 12.420 | 1560393 | 10.290 | 11801611 |
| 13.400 | 336301 | 12.800 | 1816831 | 11.950 | 25749776 |
| 14.270 | 769527 | 13.810 | 2498762 | 13.960 | 46415740 |
| 14.500 | 919077 | 14.000 | 2603640 | 16.360 | 72353576 |
| 15.320 | 1452253 | 14.500 | 2930895 | 19.240 | 104084552 |
| 16.570 | 2301906 | 15.000 | 3258405 | 22.700 | 142454688 |
| 18.080 | 3337804 | 16.000 | 3912660 | 26.840 | 188495280 |

Last but not least, the net flows going and coming to/from each wetland water body were calculated. A snapshot of the discharge exchange with the Main Abras can be seen in Fig. IV.5-4. Positive discharge values indicate flow to the wetland, otherwise flow from the wetland. The initial stage value was similar to the weir's top elevation (10.10m) to simulate the dry conditions of December, just before the rainy season. There are clear interactions between the two water bodies: the inflows and outflows can reach as high as 75 and -55 $m^3$/s respectively. During May there is a clear decline in the stages as the end of the wet season approaches; however there are two clear spikes in early June and November as mentioned before. Besides, in these two situations, the equilibrium is rapidly reached as the rest of the dry season the wetland water surface remains almost stagnant, with velocities close to zero and therefore a null flow exchange.

Table IV.5-2 Weir data for the Abras water bodies.

| Wetland | River | Station (m) | weir width (m) | Length (m) | Computations | Weir coefficient |
|---------|-------|-------------|----------------|------------|--------------|------------------|
| West Abras | Nuevo | 22499 | 33 | 15 | Standard Weir equation | 0.7 |
| Central Abras | Nuevo | 19540 | 10 | 13 | Standard Weir equation | 0.7 |
| Main Abras | Nuevo | 15220 | 90 | 440 | Standard Weir equation | 0.7 |

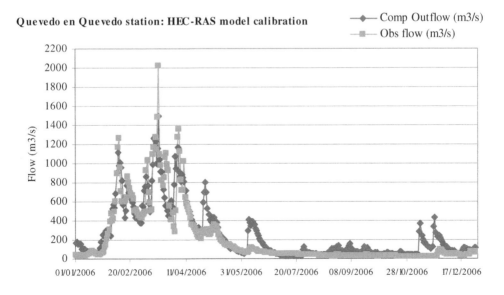

Fig. IV.5-3 Hydrograph comparison for the HEC-RAS model.

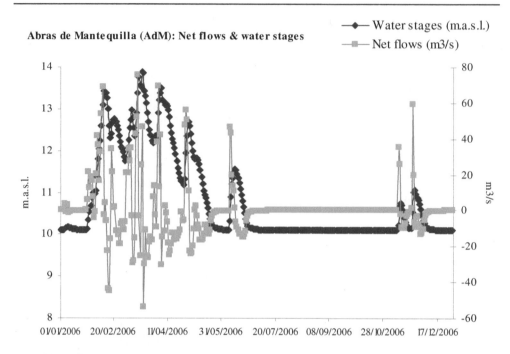

Fig. IV.5-4 Flow exchange and stages between the Nuevo River and Main Abras.

## IV.6  The water allocation model (Guayas River Basin).

### IV.6.1  Setup and simulation of current conditions

Having in mind the simulation of the current conditions, future scenario(s) and possible management solutions, a water allocation model was conceived. WEAP (SEI, 2009) is a conceptual modeling software, developed by the Stockholm Environmental Institute to represent flow distribution schemes across watersheds. WEAP uses elements such as rivers, reservoirs and water diversions to represent water flows and storages, whereas cities and crops entail water demands, everything with a monthly or yearly time step.

The water distribution model for a part of the Guayas River Basin is shown in Fig. IV.6-1. The starting year for analysis was 2007. In addition, this year was the starting point of the Baba hydropower project (to be described later on), therefore it was an ideal status to simulate the *natural conditions*. Herewith some important features of the model:

•   The Daule River with three connected water transfer projects and reservoirs: Daule-Peripa, La Esperanza and Chongón (currently in operation). La Esperanza and Chongón dams, located outside the Guayas River Basin (Fig. IV.2-1), demand

18 and 44 m³/s from Daule-Peripa, respectively, during the dry season (CEDEGE, 2008).

- The Vinces River and its towns along (e.g. Quevedo, Balzar and Vinces), from the junction between the Baba & Toachi rivers up to Vinces town downstream area.

- Crops next to the rivers (e.g. corn, rice, african palm).

- Lateral flows from Lulu and San Pablo Rivers to the Vinces (coming from the HEC-HMS model).

- The Chojampe River from its various headflows up to the AdM. The wetlands (west, central and main Abras) were simulated as a "reservoir", connected with the former and the Nuevo River.

- Water delivery losses (e.g. consumption, evaporation, infiltration).

- Groundwater flow from/to the aquifer beneath the wetland, where conductivity was estimated in 23 m/day, with extraction potential as high as 40 l/s, according to recent field campaigns in the area between the Vinces river and AdM (Romero et al., 2009).

- Two major infrastructure works (evaluated as sub-scenarios): Baba Dam (Efficacitas, 2006) on the Vinces River and the DauVin project (ACOTECNIC, 2010).

- Available measurements of nitrites, nitrates and phosphates (van Griensven and Alvarez-Mieles, 2009) were also introduced in WEAP. A mass balance for each parameter was computed along the wetland branches with emphasis in the intersection with the Nuevo River. That was the basis to compute the Water Quality Index (Brown et al., 1970; Oram and Alcock, 2010).

Water demands from crops were assumed be in principle not satisfied during the dry season (May to December). These values are shown in Table IV.6-1. To compute the total volume demanded by each crop, water demands were multiplied by the landuse areas retrieved from local landuse maps (CEDEGE, 2002) and satellite imagery (Fig. II.3-2). For most of them, the only available information provided combined uses (e.g. 70% cultivated grasslands and 30% fruits, or short-term crops). Therefore, previously it was necessary to calculate the composed demand according to the individual consumption and the surface percentage on the map.

Furthermore, for the most important cities namely Quevedo, Vinces and Daule, the population was estimated based on the last available census (INEC, 2002). These estimations were influenced by the percentage of water accessibility, 77 % in the Guayas River Basin and 44 % in the wetland area (Debels et al., 2009), and establishing a standard use of 150 liters/person/day. In addition, water losses after consumption in cities ranged between 25 and 50% and in crops were as high as 65% (Villa-Cox et al., 2011). Nonetheless, the latter was fixed in 54% for most of the crops and 30% for the natural pastures when combined with forestry. Finally, the priorities (1 for the most urgent and 99 for the least important) were assigned as 1 for cities, 2 for crops and >3 for cultivated grasslands.

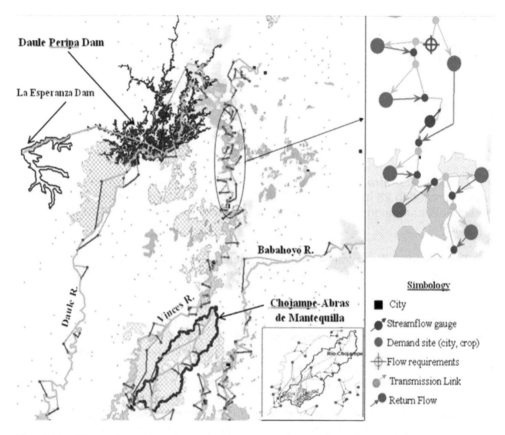

Fig. IV.6-1 Water allocation scheme for the central Guayas River Basin.

As it was pointed out before, there are some hydraulic works in the Guayas River Basin, currently in operation. The project includes a water diversion of 5 km$^3$/year out of the basin, towards La Esperanza Dam (volume = 450 Hm$^3$) in the Manabí Province (Fig. IV.2-1). To represent in WEAP this and other similar diversions, from the source reservoir a diversion canal was linked to a "natural river" and this latter was connected to the target reservoir (Fig. IV.6-1). Moreover, in the lower course of the Daule River there is another important diversion to the Santa Elena Peninsula, via the Chongón Dam (280 Hm$^3$ of volume), with irrigation purposes (CEDEGE, 2008). Data from the Baba project were not included in the baseline analysis but on the corresponding scenario, described shortly below.

Table IV.6-1 Crop water demands in AdM and along its adjacent rivers.

| CROPS | Water consumption, $m^3$ / Ha | | | | | | | | |
|---|---|---|---|---|---|---|---|---|---|
| | May | Jun | Jul | Aug | Sep | Oct | Nov | Dec | Total |
| short-term crops | 396 | 689 | 840 | 742 | 133 | 133 | 133 | 133 | 3200 |
| rice | 384 | 0 | 993 | 1121 | 1161 | 1139 | 536 | 0 | 5333 |
| 70% rice, 30 % cultiv grasslands | 381 | 113 | 807 | 897 | 925 | 910 | 487 | 113 | 4633 |
| 70% natural grassl, 30% rice | 115 | 0 | 298 | 336 | 348 | 342 | 161 | 0 | 1600 |
| 50% rice, 50% cultiv grasslands | 380 | 188 | 684 | 748 | 768 | 757 | 455 | 188 | 4167 |
| cocoa | 906 | 913 | 922 | 996 | 964 | 896 | 749 | 320 | 6667 |
| banana | 862 | 1027 | 1164 | 1316 | 1126 | 1024 | 993 | 488 | 8000 |
| 70% banana, 30% cultiv grassl | 716 | 831 | 928 | 1034 | 901 | 829 | 808 | 454 | 6500 |
| 50% banana, 50% short-term crops | 629 | 858 | 1002 | 1029 | 630 | 579 | 563 | 311 | 5600 |
| 70% coffee, 30% cocoa | 972 | 974 | 977 | 999 | 989 | 969 | 925 | 796 | 7600 |
| sugar cane | 0 | 471 | 716 | 964 | 986 | 726 | 0 | 0 | 3862 |
| african palm | 1250 | 1250 | 1250 | 1250 | 1250 | 1250 | 1250 | 1250 | 10000 |
| corn | 329 | 694 | 883 | 761 | 0 | 0 | 0 | 0 | 2667 |
| 70% corn, 30% rice | 345 | 486 | 916 | 869 | 348 | 342 | 161 | 0 | 3467 |
| 50% cultiv grassl, 50% nat grassl | 188 | 188 | 188 | 188 | 188 | 188 | 188 | 188 | 1500 |
| 50% cultiv grassl, 50% fruits | 604 | 604 | 604 | 604 | 604 | 604 | 604 | 604 | 4833 |
| 70% cultiv grassl, 30% banana | 521 | 570 | 612 | 657 | 600 | 570 | 560 | 409 | 4500 |
| 70% cultiv grassl, 30% fruits | 513 | 513 | 513 | 513 | 513 | 513 | 513 | 513 | 4100 |
| 70% cultiv grassl, 30% nat. forests | 263 | 263 | 263 | 263 | 263 | 263 | 263 | 263 | 2100 |
| 70% cultiv grassl, 30% rice | 378 | 263 | 560 | 599 | 611 | 604 | 423 | 263 | 3700 |

WEAP also included the three wetland water bodies described earlier (Fig. IV.5-2). With this aim, the available DEM was a source for the elevation-volume curves for each wetland (as for the HEC-RAS model, Table IV.5-1). Additionally, the volumes of capacity (maximum volume), "inactivity" (minimum operative), initial (condition) and "conservation" (e.g. for floods or irrigation) were included during the model setup as well as for all the existent and projected reservoirs (CEDEGE, 2008). The Table IV.6-2 presents a summary of these data. Ecological flow were adopted based on earlier estimations along the Daule and Vinces River (ACOTECNIC, 2010; Efficacitas, 2006). These values were 50 and 10 $m^3$/s, respectively.

After the baseline analysis was carried out, some important results were obtained. First of all, the volumes belonging to the wetland reservoir (95 to 230 $Mm^3$/year) can be seen along the Chojampe River (Fig. IV.6-2). Secondly, along the Vinces River case (Fig. IV.6-3), the important contribution from the Lulu & San Pablo Rivers could be quantified. The Vinces' discharges increased in 45% after the union with the Lulu River and up to 163% after the union with the San Pablo River. Thirdly, an important streamflow is naturally diverted to the Nuevo River, around 32-33% of the Vinces' flow.

Table IV.6-2 Data of projected, existent and natural reservoirs.

| Reservoir | River | Volumes (hm$^3$) | | | |
| --- | --- | --- | --- | --- | --- |
| | | Capacity | Initial | Conservation | Inactivity threshold |
| Daule-Peripa | Daule | 6000.0 | 6000.0 | 5500.0 | 4000.0 |
| La Esperanza | La Esperanza | 450.0 | 400.0 | 450.0 | 93.0 |
| Chongón | Chongón | 304.0 | 270.0 | 280.0 | 50.0 |
| Main Abras | Chojampe | 188.5 | 8.0 | 186.0 | 8.2 |
| West Abras | Nuevo | 3.3 | 0.1 | 3.3 | 0.9 |
| Central Abras | Nuevo | 2.5 | 0.9 | 2.5 | 1.8 |
| Baba | Vinces | 138.0 | 93.0 | 93.0 | 17.8 |

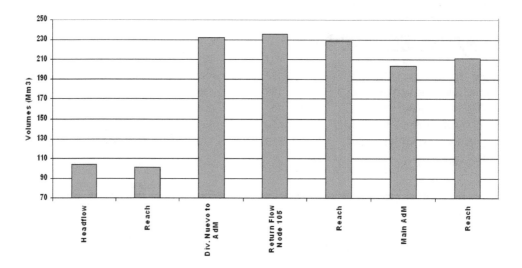

Fig. IV.6-2 Annual volumes (Mm$^3$) along the Chojampe River, current conditions.

The Nuevo River receives the heaviest influence from the main Abras, in comparison with the other two water bodies (west and central). This occurs with higher evidence from February to April, where the discharge differences are around 14 to 21 m$^3$/s (Fig. IV.6-4), comparable with most of the inflows/outflows calculated with HEC-RAS (Fig. IV.5-4). This exchange is minimal from May to October, which is coherent to what has been observed every year. Finally, the discharges on the Daule River are shown in Fig. IV.6-5. A small decrement as consequence of the diversion to La Esperanza Dam (around 8%) was observed as well as small diversions due to crop production along the river (mainly rice and in general short-term crops), ending with another important decrement of around 20% due to the demand from the Chongón reservoir.

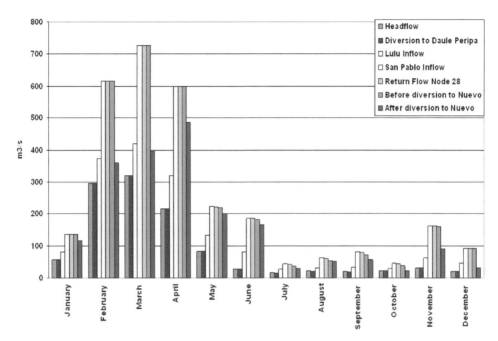

Fig. IV.6-3 Monthly flows along the Vinces River, current conditions.

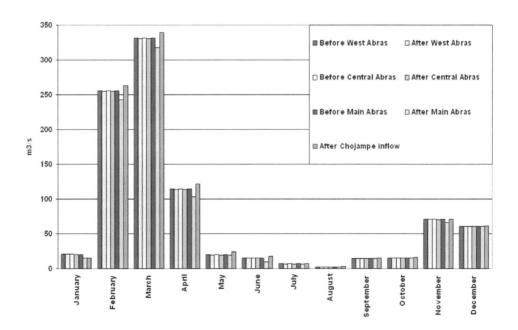

Fig. IV.6-4 Monthly flows along the Nuevo River, current conditions.

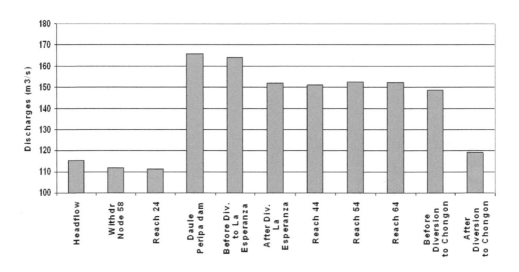

Fig. IV.6-5 Annual average flows along the Daule River, current conditions.

## IV.6.2  Simulation of Scenarios

First of all, it was investigated the exclusive influence of the Baba project upon the Vinces River (Fig. IV.6-6). Initially the river had an average discharge around 92 m$^3$/s. However, after the water transfer to the Daule-Peripa dam was incorporated into the analysis, the river flow declined to 21 m$^3$/s (i.e. -75%). Actually, this might be the most critical segment of the river. Fortunately, there was some recovery after the inflow from the Lulu (Q increases to 63 m$^3$/s) and the San Pablo River (up to 174 m$^3$/s) showing however still a drop of around 29-30%. Yet downstream this reduction remained likewise in both the lower course of the Vinces River and the natural diversion to the Nuevo River.

Due to the climatic tendency discussed in section II.4.1, the selected case for the Guayas River Basin was a temperature increment by +0.5°C (the closest to the expected one) for the period 2007-2043. The projected rainfall time series (provided by PIK) were inserted in each of the stations of the HEC-HMS models (built initially for the current conditions) producing thus new headflow and lateral flow values for each river in the water allocation scheme. The time span was divided in four sub-periods to facilitate the calculations in WEAP:

• Representative year A: period 2007-2010 (1st "decade")

• Representative year B: period 2011-2020

- Representative year C:  period 2021-2030

- Representative year D:  period 2031-2040

The Baba Dam was included in the climatic scenario analysis because it is expected to have it fully operational during the second half of the $2^{nd}$ decade onwards if not before.  The flow distribution across the links between the Nuevo River and the wetlands (how much in percentage is diverted to/from each wetland body) remained similar to the current conditions.  This was because climate changes strictly supposed only a variation on the subbasin headflows and not water diversions along the rivers or tributaries, for instance. Nonetheless, the magnitude of the streamflow did vary remarkably along the Nuevo River. First, it decreased in 30% due to the influence of the dam and then, following the patterns described by Nieto el al. (2002), the average annual flows recover, in particular during the year type C (decade 2020-2030), around 9-10% and in the fourth decade up to 30% (Fig. IV.6-7), always compared against the base year (2007). As a corollary, there was a general recovery on the flow regime after the $2^{nd}$ decade.  Since later on the proposed management options will be compared against this composed new baseline scenario, all comparisons will be carried out taking in consideration only the two latest decades.

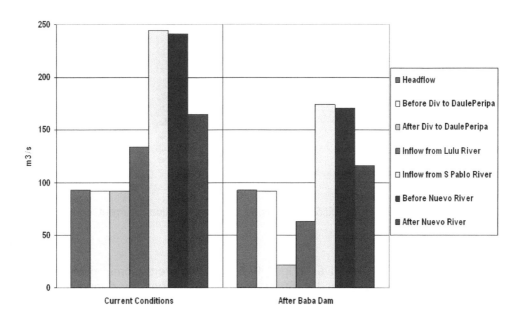

Fig.  IV.6-6 Average annual flow along the Vinces River once the Baba Dam becomes operational ($2^{nd}$ decade).

Given the imminence of the construction of the DauVin project, it was necessary to evaluate the potential influences on the system. According to the planned water distribution scheme (Fig. II.4-2) the initial concern was to determine whether the project restrain some water from the wetland, since in the available literature there was no mention of AdM at all (ACOTECNIC, 2010). It was observed that there were volume increments (around 13%) but only for the Nuevo River and after the junction with the Chojampe River (red ellipse in Fig. IV.6-8). As expected, there were decrements (7%) in the downstream area, consequence of the water transfer to the Puebloviejo River. It was found in literature that in fact the 25.5 m$^3$/s entering the Nuevo River area flow along an old path, south of the current one, siphoning the river itself. Hence, the upstream part of the Nuevo River remains untouched with neither positive nor negative implications from the mentioned project.

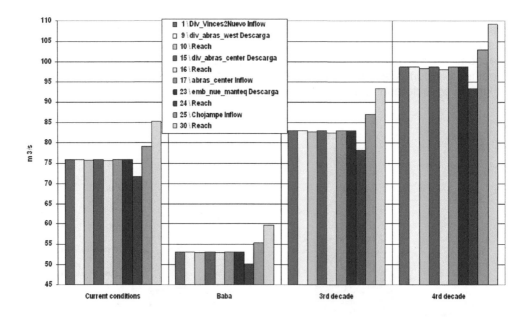

Fig. IV.6-7 Annual average discharges along the Nuevo River, current conditions vs. scenarios.

Finally, along the Daule River (Fig. IV.6-9) there were some river reaches downstream with flows under the 50 m$^3$/s threshold, especially during the driest months (August, September, around only 8 m$^3$/s after the Chongón diversion). This may occur even counting on the water transfer from the Baba Dam to Daule-Peripa. Consequently, a serious danger to the water intake of the Guayaquil city might be caused, not only in water quantity, but in water quality due to probable higher concentrations of nutrients and other substances.

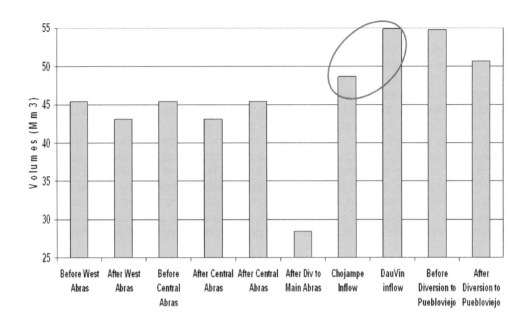

Fig. IV.6-8 Effect of the DauVin Project on the Nuevo River.

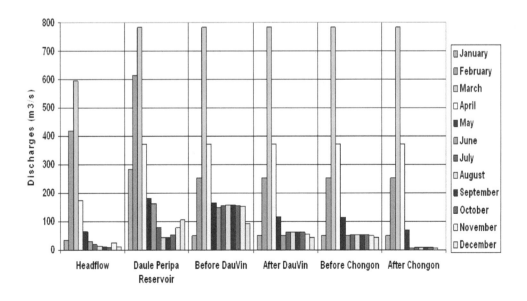

Fig. IV.6-9 Flows along the Daule River and the DauVin project.

## IV.6.3 Simulation of Management Solutions

A comparative chart on the performance of the management solutions in the wetland area is shown in Fig. IV.6-10. In general, remarkable increments in water quantity (+28%) can be observed between MS0 (BAU) and MS1 when the use of gates was introduced. In MS1, taking the baseline as reference, the inactivity level (minimum level) at the AdM *reservoir* was fixed at 13.96 m, which implied around 4 meters above the top level of the existent natural weir. The increment in water storage via the reduction of discharges from the Chojampe River to the Nuevo River was around 23 $Hm^3$/year. When the climate conditions were inserted, the increased volumes were estimated as 70 and 93 $Hm^3$/year in 2020-2030 and 2030-2040, respectively, should the elevation be increased even more. In addition, the volumes that this enlarged *reservoir* could release to the Nuevo River might also increase, in 11% per year with no climatic variation (e.g. 2nd decade) and 33 – 44 % with an increment of rainfall and corresponding discharges (3rd and 4th decades).

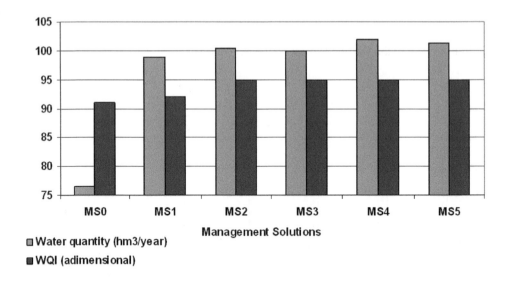

**Management Solutions: score on quantitative indicators**

Fig. IV.6-10 Performance of MS in the wetland area (Water quantity & quality).

Because the order of magnitude of water retentions and diversions is larger than the one originated by changes in landuse, there was a reduction in the rising rate of water quantity for the rest of options (MS2 to MS5). It is noteworthy that even more aggressive landuse substitutions (MS3 vs. MS2) not necessarily entail improvement in water quantity; in fact, long-term crops (e.g. cocoa) also require massive amounts of water. Therefore, MS2 was better (and possibly less costly) than MS3. A similar result was observed when comparing MS4 (MS2 + reforestation) with MS5 (MS3 + reforestation), for MS5 should be expected more expensive in terms of time and money than MS4.

Finally, the Water Quality Index (WQI) methodology (Brown et al., 1970) was adopted to establish acceptance levels in the wetland area. In general, WQI levels are very high and still are in line with a previous report (Prado et al., 2004). Unlike the case of water quantity, more noticeable improvements are observed not when water is retained but when better agricultural practices are introduced and even more (3%, MS1 to MS2) when those are combined with crop substitution. Thenceforth, the situation remains more or less similar (WQI around 95) for the rest of the options, even MS4 and MS5, perhaps due to an insufficient reforestation rate (5% per decade). At the end, MS4 emerges, from the hydrological point of view, as the best alternative to improve the two quantitative indicators in the system.

## IV.7 Value functions for the quantitative indicators.

The decision support system (Chapter VI), one of the components of this thesis, demands information from the quantitative indicators. However, across the set of indicators, magnitudes and units are not the same evidently. In order to fairly compare these results with other sort of parameters (e.g. socio-economic ones, Chapter V), model outcomes must be translated into a standardize *scale*. This procedure was carried out via the so-called value functions, of course, one for each variable or indicator.

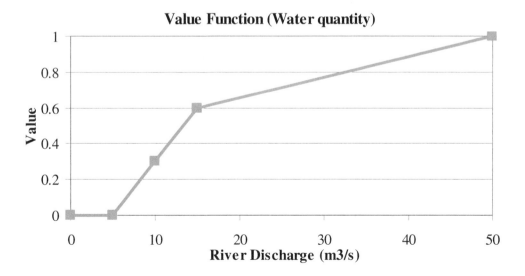

Fig. IV.7-1 Adopted value function for water quantity.

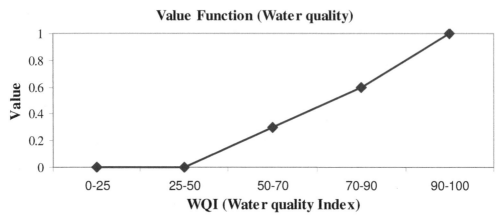

Fig. IV.7-2 Adopted value function for water quality (Oram and Alcock, 2010).

For water quantity a minimum threshold of 5 m$^3$/s was selected, based on ecological flows in the Vinces River (10 m$^3$/s), suggested by previous environmental impact assessments (Efficacitas, 2006). Moreover, for navigability purposes and irrigation in riparian areas and in absence of better judgement, a discharge of 10 m$^3$/s was assumed equivalent to 0.3. Furthermore, for flows up to 15 m$^3$/s, meaning irrigation for larger riparian areas, the function value was 0.6. Finally, the top value is reached when Q is higher or equal to 50 m$^3$/s. Finally, between the mentioned values, the function pattern was assumed linear (Fig. IV.7-1). On the other hand, for water quality, a field manual for water quality monitoring from the US National Sanitation Foundation was used to build the value function depicted in Fig. IV.7-2 (Oram and Alcock, 2010).

## IV.8 Concluding remarks

A modeling framework has been proposed for the analysis of scenarios and management options in the Vinces catchment and the Abras de Mantequilla wetland. Basically, it has two main functions: serve as a tool to characterize the system and as a data carrier to the wetland area where information is scarce. In addition, this system of models responded to the previously identified DPSIR chains for the present case study (Zsuffa and Cools, 2011). The two main drivers were the climatic variations and the major infrastructure works (catchment scale) planned by SENAGUA, namely the Baba Dam and the DauVin project. As a consequence, two main indicators were assessed, from a modeling perspective: water quantity and quality.

From the rainfall-runoff model in the Vinces catchment we could learn some lessons. Firstly, despite the strong assumptions that were made the results could be acceptably verified in several locations, for instance, at the Baba, Toachi, Pilalo catchment outlets and the *Quevedo at Quevedo* river station. In fact, having reacted to a certain precipitation input, the model sometimes *surprised* the corresponded observation and thus the latter became somehow *suspicious*. This happened mainly during some days in June and November (during the dry season) as consequence of localized stormy events in the Andes. Secondly, there was an important flow contribution from the San Pablo and Lulu Rivers which in a nearby future may attenuate the flow reduction foreseen by the Baba Dam along the Vinces River. This confirms recent conclusions on the matter (Efficacitas, 2006).

Satellite data, specifically TRMM, proved to be a suitable complementary precipitation data source for the Vinces upper catchment. The proposed methodology followed current approaches such as the typical comparison against the ground-based data. It also showed an easy yet effective way of correcting the monthly bias of TRMM data at a catchment scale. By making use of the rainfall time series from raingauges, the bias-corrected monthly TRMM data were disaggregated to a daily resolution. Such procedure was necessary to use the satellite borne precipitation in combination with the in-situ data as input for the HMS model. In spite of some slight overall reductions on the model efficiency, possibly due to the empirical temporal disaggregation, this new simulation showed outcomes very comparable with those using only raingauge information. Therefore, the contribution of this approach was based on the fact that, given a known bias, the satellite data could still be corrected and yet may resemble the information provided by the ground stations.

Another simple hydrologic model was setup for the Chojampe catchment, where the wetland is located. Following similar methodology as in the Vinces model, its main goals were to provide the boundary conditions for the water allocation model and thus help characterizing this almost ungauged catchment. An important drawback was that some important physical aspects such as the groundwater connection with the river were not covered in detail by the HMS model due to constraints in data coverage. As an alternative, calibration was a key resource to determine optimized parameters for each micro-catchment.

Along the Vinces and Nuevo Rivers as well as the AdM a hydrodynamic simulation was built up. Yet another data carrier, the HEC-RAS model used the flows from Lulu & San Pablo Rivers as part of its boundary conditions and then computed flows and water stages in the downstream area. The flow values between the Nuevo River and the Abras were very high during the rainy season but in May commonly turned to a stagnant level once the precipitation regime ended. Lastly, there was a noticeable exchange in early June and November of course as consequence of the isolated stormy events in the Andes that were routed throughout the model cascade.

A water allocation model computed the distribution of volumes across the main rivers of the Guayas watershed. The expected increased climatic variations, the major hydraulic works as well as the proposed management options were assessed using this tool. Firstly, it was seen that the flows along the rivers may tend to increase noticeably after the second decade (i.e. after 2021) consequence of the incremental trend in precipitation, expected for the Ecuadorian Coastal Region. Secondly, it was confirmed the positive influence the Lulu & San Pablo Rivers have on the system, in particular once the Baba Dam will enter in operation and even more when the climate changes take place. As a result, the effects on the wetland might be fully attenuated as the time span approaches to 4 decades. Moreover, the DauVin project has, at least for the Vinces River and the wetlands, no noticeable influence, being the sole positive effect when a residual inflow enters the Nuevo River but for its downstream area. That is not the case for the Daule River when a critical water shortage for Guayaquil City might be expected should this water transfer be executed simultaneously with the already important water diversions in operation, i.e. La Esperanza and Chongón.

Five management choices have been proposed, all compared with the *Do nothing alternative* ("Business As Usual" -BaU). In general, for water quantity, the magnitude of variation that crop substitutions originate is small compared to direct water retention or a water transfer. The use of gates during the dry season increased remarkably the retained volume in the wetland (MS1 vs MS0). Although in lesser degree, when the MS2 was evaluated water quantity in the wetland increased noticeably as well. Thenceforth, the water volumes reduced their rate of increment but yet the water quantity showed a slight improvement when migrating from substitution alternatives (MS2 & MS3) to more elaborated alternatives that involve reforestation (MS4 & MS5). At the end, by a little difference, MS4 seems to be the *best* choice. For the WQI, the overall improvements are less noticeable (percentages less than 5%) due to two reasons. The wetland is still relatively in good condition (WQI is very high), as reported before (Prado et al., 2004). Secondly, crop substitution, when aggressively implemented (MS3) not necessarily implied higher WQI than MS2. It seems then that increasing the rate of substitution may not be worthwhile in terms of greatly improving the index, which is already quite high. Similar situation is seen between MS4 & MS5. Hence, the WQI was not that sensitive when moving from less to more complex management options.

Ultimately, an integrated analysis cannot be based exclusively on quantitative indicators. Firstly, insufficient data leads to the necessity of expert judgment or *human feedback*. Secondly, the inherent nature of socio-economic indicators makes the opinion of stakeholders and experts crucial to complement the hydrological perspective that the management solutions (e.g. MS2 or MS5) initially showed. Hence, a more clarification on a more realistic rank of solutions was still required.

$$\mathcal{C}hapter\ \mathcal{F}ive$$

# V  Stakeholder appreciation of management solutions*

## V.1  Introduction

In Chapter IV a modeling framework was implemented where the water quantity and quality indicators were assessed for both a baseline scenario and each of the proposed management solutions. Clearly, not every indicator can be determined by computer-based simulation, mainly due to lack of numerical data (time series, spatial maps, etc), knowledge or experience. In general, a severe problem of data scarcity is a common situation when biological, ecological as well as socio-economic and even hydrological data are required; therefore, there is an urgent need to seek other sources of information, for instance, human expertise. To this end, expert elicitation has been used in this research.

Expert elicitation is not a new data collection technique. In some past and recent studies involving wetlands, riparian areas and other water systems, workshops were held involving not only academic experts but also public in general and particularly local stakeholders (Reckhow, Arhonditsis et al. 2005; Swor and Canter 2008; Titus, Hudgens et al. 2009; Cleland and McCartnety 2010). However, the wetland dwellers may qualify also as *experts* because (i) they know very well the study area; (ii) they have witnessed, in some cases for decades its evolution; (iii) they may have simple but interesting views on what could be suitable solutions to the current problems; and (iv) they can be key members for implementing some solutions in field. This particular sort of data collection is crucial to refine the attributes, a key concept in the elaboration of a Decision Support System (Chapter VI).

* Parts of this chapter are based on: Arias-Hidalgo, M., Villa-Cox, G., van Griensven, A., Solórzano, G., Villa-Cox, R., Mynett, A.E., Debels, P. (2012). "A decision framework for wetland management in a river basin context: the "Abras de Mantequilla" case study in the Guayas River Basin, Ecuador", Journal of Environmental Science & Policy, Sp. Ed. (accepted for publication), doi:10.1016/j.envsci.2012.10.009.

## V.2 Expert elicitation in AdM.

The socioeconomic implications that may take place once the management options are applied on the system were evaluated. Differently from indicators such as water quantity and quality, there were no local-scaled models that simulated the dynamics that socioeconomic variables may have shown under the various management strategies. Despite this initial setback, it was still possible to determine the present value of such indicators and the potential response under the BAU scenario. In fact, there was some available information about landuse as well as the socioeconomic situation of the wetland dwellers (INEC, 2002).

An alternative to tackle severe data scarcity is expert elicitation. An example of this would be to obtain the opinion about the variation of costs of each option under the different choices. Using interviews or workshops, it is possible to ask one or more local scholars or experienced people (10 people) to score on a predefined scale what would be the consequences, either positive or negative (or sometimes none) of each MS with reference to a certain variable or indicator.

Table V.2-1: An example of the Lickert scale (economic indicator: reduction of crop costs).

| Category | Description | Lickert scale, $a_i$ |
|---|---|---|
| Remarkable improvement | Significant costs reduction | 5 |
| Moderate improvement | Small costs reduction | 4 |
| Indifferent to BAU | Same as current cost structure | 3 |
| Moderate detriment | Small increment in costs | 2 |
| Significant detriment | Significant increment in costs | 1 |

A convenient ranking for this methodology can be found in the centered Lickert scale (Burns and Bush, 2008). The list goes from 1 to 5. A value equal to 3 implies the management solution does neither enhance nor worsen the current status. Values higher than 3 (4 & 5) indicate minor or major improvements referred to the BAU; therefore the MS could be considered as *recommendable*. Conversely, scores lower than 3 entail bad or worse consequences (2 & 1, respectively) for an indicator when applying the MS, always in comparison with the baseline. Table V.2-1 shows an example on a *reduction of crop costs*. After a management solution is applied, an increment of this variable would mean an improvement in the situation and viceversa. Thus the scale permits a qualitative measurement of the impact of the evaluated alternative, according to the expert criterion. In case the option might not be relevant to the indicator or does not have any relation whatsoever, the expert could still score a "3", thus not affecting at all further outcomes.

It must be recalled that in most of the indicators/fields there will be more than one consultant. For those cases, it was necessary to find a method that represents all opinions in a fair way. A simple technique in this regard is to average the scholars' scores for each indicator. Nevertheless, some assumptions must be taken in account before applying it:

•   There are no *a priori* reasons to think that the opinion of one expert have more weight than others.
•   The selection must be done carefully to avoid *careless or wrong* assessments that influence negatively the final outcome of the indicator.
•   The resultant average value for each indicator is within in the same interval as the Lickert scale (1 to 5).

## V.3  Socio-economic / Institutional indicators and value functions

Several socioeconomic variables have been considered for the qualitative approach. Unfortunately, for these indicators there were not enough numerical data or existent models to quantify either a current status or probable projections. Nonetheless, their relevance when assessing the performance of each management solution should not be overlooked. Some of the main involved indicators (mostly define by their name itself) were:

•   Crop investment / sowing costs, per Ha: Mainly related to a probable increment /decrement of crop costs at the initial stages of a management solution (MS).
•   Crop maintenance costs, per Ha: In the aftermath of the application of a MS.
•   Food security: How feasible are the food sources for the local stakeholders?
•   Crop productivity per Ha: How helpful a MS can be to enhance the crop productivity?
•   Income level of local population: Does the evaluated MS help increasing the revenues for the dwellers?
•   Economical associability of local stakeholders: Does the assessed MS promote or damage the potential of association of the wetland inhabitants?

And so forth. A complete list of the indicators and their categories is provided in Appendix B.

With the aim to convert the indicator number to a normalized function whose output falls within the [0,1] interval, it was necessary to apply a value function (Duckstein and Goicoechea, 1994). This is a requirement similar to the one carried out for the quantitative indicators (Chapter IV) prior to the calculation of the evaluation matrix for the DSS setup (Chapter VI).

For the socio-economic variables, four value functions were considered (Fig. V.3-1). The linear type implies that weights in the qualitative scale are equally important in terms of the impact caused by the specific criterion on the decision. Secondly, the Quadratic function means that the rate of impacts increases (steeper slope) as the indicator values grow. Conversely, the (1+log) suggests that the lower the values the greater the rate of impacts. Finally, the power function, defined as follows:

$$Value = \frac{a^n}{a}$$

Eq. V-1

where $a$ is an integer number, and $n$ is the Lickert's score. This function behaves similar to the quadratic function, but assigning higher scores to the lower indicator values and showing milder impact slopes for higher indicator values. All these options were considered in the decision support tool (Chapter VI). Similar and alternative value functions are reported in the literature (Duckstein and Goicoechea, 1994).

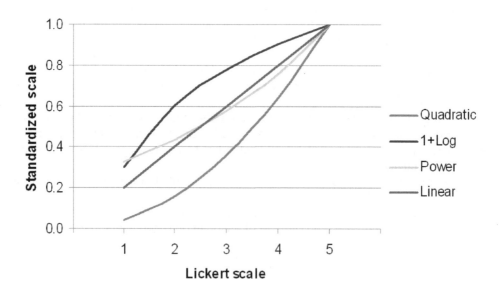

Fig. V.3-1 Comparative chart of several value functions for qualitative indicators

With respect to the institutional indicators, a similar procedure was envisaged but involving some changes. The aim was not to compare the status quo vs. the impact of a certain MS, but to measure the institutional capacity (in its different levels) to adopt the proposed alternative. In Table V.3-1 there is an example of a Lickert scale for an institutional indicator (*local management capacity*). The scale is the same as in other indicators keeping thus consistency when applying the value functions and evaluating thus all variables in the same context. A more detailed list of the institutional indicators, with their Lickert values and descriptions is shown in the Appendix B. Each indicator has a code or abbreviation to facilitate the further implementation of the DSS process.

Table V.3-1 Lickert scale for the local management capacity.

| Category | Description | Lickert scale, $a_i$ |
|---|---|---|
| Optimal | Management and coordination capacity of the regional management structure is excellent to adopt the alternative. | 5 |
| Adequate | Management and coordination capacity of the regional management structure can appropriately face the measure, however still prone for enhancement. | 4 |
| Acceptable | Management and coordination capacity of the regional management structure can still face the measure, but requires strengthen to do it adequately. | 3 |
| Deficient | Management and coordination capacity of the regional management structure is deficient; requires major changes to adopt the alternative. | 2 |
| Not suitable | There might not be sufficient management and coordination capacity to implement the alternative and it is not possible to do improvements. | 1 |

## V.4  Ecological indicators and value functions

### V.4.1  Biodiversity

The Shannon & Wiener index (Hill, 1973; Krebs, 1978) is a way to measure biodiversity. This coefficient combines (i) the number of species and (ii) the equality or inequality of the individuals distribution of different species (Krebs, 1978; Lloyd and Ghelardi, 1964). Hence, the larger the number of species or a more uniform distribution of species, the larger the value of the index. The index is expressed as follows:

$$H' = -\sum_{i=1}^{S} p_i \log_2 p_i$$

Eq. V-2

where:

S = Number of species (the richness);
$p_i$ = Ratio between the individuals of the species and the total number of individuals, i.e. the relative abundance of the especies = $n_i/N$;
$n_i$ = Number of individual of species i;
N = Total number of individual of all species;

The qualitative analysis for the biodiversity indicator (code name: *biodiver*) was very similar as for the previous types of indicators. The suggested minimum value in the literature for this index is 1.5 whereas the maximum for sustainable ecosystems is 3.5, rarely exceeding 4.5 (Magurran, 1983; Margalef, 1972). With this in mind, the Lickert scale is proposed in the following way (Table V.4-1). In the wetland, the current observed average index value was 2.5 and this may correspond to category 3, i.e. indifference with respect to the BAU. A potential loss of biodiversity may have to consequences on the system. One of these would be changes in the ecosystem functioning and the loss of buffering capacity.

Table V.4-1 Lickert scale and value function for the biodiversity indicator.

| Category | Impact description | Lickert Scale $a_i$ | Value function $f(a_i)$ |
|---|---|---|---|
| Significant improvement | Significant increment on biodiversity | 5 | 1.00 |
| Small improvement | Small increment on biodiversity | 4 | 0.64 |
| Indifferent to BAU | Current level of biodiversity | 3 | 0.36 |
| Small detriment | Slight loss of biodiversity | 2 | 0.16 |
| Significant detriment | Significant loss of biodiversity | 1 | 0.04 |

## V.4.2  Degradation

The habitat degradation (code: *degrad*) is a key cause of the loss of biodiversity. Higher losses are a consequence of large area reductions. Anthropogenic activities such as biomass extractions from the ecosystems for fodder purposes led to an important declining on soil protection and increment of its vulnerability against the erosive action of climatic agents. This has been seen in the wetland (especially on the eastern part). These events also reduce the soil fertility and water storage capacity. Agricultural practices and land degradation were part of the pressures identified in the DPSIR chains (Chapter II.3). As a consequence, the natural resources (flora and fauna) have been greatly affected and in many cases lost their natural balance.

Table V.4-2 Impact categories to evaluate the vegetation integrity status within disturbance classes.

| Impact category | Description | Impact score intensity | Value function |
|---|---|---|---|
| Lower limit | - | 0.0 | 1.00 |
| None | Vegetation composition is completely natural. | 0.5 | 0.95 |
| Slight | A slight change in the vegetation composition is noticeable | 1.5 | 0.85 |
| Moderate | The flora composition has been moderately modified; however, external species are clearly less abundant than native species. | 3.0 | 0.7 |
| Considerable | The flora composition has been moderately modified although the external and/or introduced species are approximately as abundant as the native species. | 5.0 | 0.5 |
| Serious | The flora composition has been dramatically modified. However there are still some natural species although most of the vegetation consists in introduced or external species. | 7.0 | 0.3 |
| Critical | Vegetation has been completely altered and there are almost none natural species | 9.0 | 0.1 |
| Upper limit | - | 10 | 0 |

For the Abras de Mantequilla wetland, the categories to assess the vegetation status within a range of existent or potential disturbances are shown in Table V.4-2. These values were taken from the WET-Health methodology (Macfarlane et al., 2008). Score ranges were assigned to each kind of impact on the vegetation, by means of normalized categories in the interval 0 to 10. Thus, the *critical* condition entails completely modified vegetation with scores between 9 and 10 and hence a value function between 0.1 and 0 respectively (Fig. V.4-1). The methodology considers vegetation as *critical* if it is severely altered giving thus a score between 7 and 9 and hence a final value between 0.3 and 0.1 and so forth, until the *no impact* category which has scores between 0 and 1.5 and thus yielding the highest value functions (0.85 to 1.00).

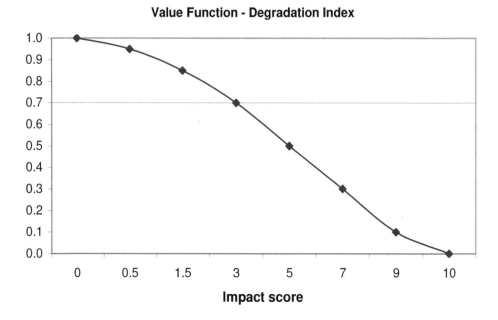

Fig. V.4-1 Value function for the degradation index.

## V.4.3  Eutrophication

When an ecosystem is massively enriched with nutrients, then it is eutrophicated. In the case of the Abras de Mantequilla wetland, there are three causing agents of eutrophication (code: *eutroph*). The first one is the households organic effluents (organic and inorganic residuals) to the water bodies. The second one is the pollution coming from agriculture and livestock activities, via inorganic fertilizers and animal excrements. The third source is the forestry pollution due to the crop residuals after agriculture, incrementing thus the dissolved organic matter in the water.

One of the first signals of eutrophication in the wetland is the rapid formation of water hyacinths (Fig. V.4-2). This phenomenon inhibits the growing of phytoplankton, reducing thus the population density of zooplankton and damaging the food chain. In addition, eutrophicated waters hamper navigability, particularly during the dry season and even sometimes during the rain period, due to their elevated reproduction rates. Other effects associated with this problem are the reduction of dissolved oxygen (and thus devastating consequences for fish populations), sedimentation increasing and a suitable habitat for mosquitoes, as vectors of tropical diseases (e.g. malaria).

The indicator for eutrophication was based on the ratio between water hyacinths coverage and the total area of the water body. This qualitative index used the same scale as the biodiversity indicator, where the medium category marked indifference with respect to the current status (Table V.4-3).

Fig. V.4-2 Water hyacinths in AdM, nearby the confluence with the Nuevo River.

Table V.4-3 Lickert scale and value function for the indicator eutrophication.

| Category | Description | Lickert Scale $a_i$ | Value Function $f(a_i)$ |
|---|---|---|---|
| Significant improvement | Significant reduction on water hyacinth coverage | 5 | 1.00 |
| Small improvement | Small reduction on water hyacinth coverage | 4 | 0.64 |
| Indifferent to BAU | Current average water hyacinth coverage | 3 | 0.36 |
| Small detriment | Small increment on water hyacinth coverage | 2 | 0.16 |
| Significant detriment | Significant increment on water hyacinth coverage | 1 | 0.04 |

## V.5  Evaluation of the Management Solutions

For the Abras de Mantequilla case study, a panel of 10 experts was consulted about the suitability of the proposed management options.  This panel, albeit in short number, covered a wide range of fields such as ecology, water management, biology, land and water development, chemistry, economy.  The consulted were staff of the following institutions:

- Ministries of Agriculture and Environment.
- Local municipalities.
- United Nations Development Program.
- Local universities.

An example of the questionnaire employed to collect the criteria from these experts is provided in Appendix A.  Before answering, the panel of experts received an inception about the current status of the system, namely the baseline and the involved scenarios as well as the different proposed management solutions (Chapters II & IV).  Afterwards, they were carefully instructed on how to score, given the categories and description of each indicator.  Should they be not specialists in a certain area, they still had the chance to score 0 and thus not load wrong weights upon the final MS ranking.

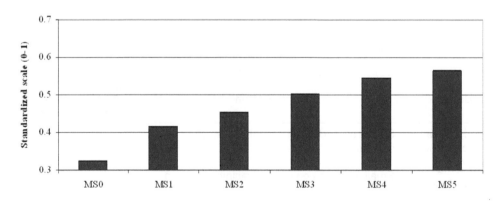

Fig. V.5-1 Ranking of Management Solutions according to expert elicitation.

In most of the indicators it seemed the MS5 & MS4 were the preferred options by the consultants. There seems to be a tendency towards food safety and biodiversity. In the former preference, the experts considered that more elaborated management solutions may provide more food sources, through a more aggressive crop substitution and reforestation. As for the second choice, they granted more weight to those alternatives dealing with the ecological corridors as potential hubs for species. Nevertheless, there were some exceptions, especially in those related with costs (e.g. Crop investment/sowing per Ha. for the MS4 & 5 may imply higher total investments to be operative than *simpler* MS, namely 1 and 2). An overall ranking summarizes what the experts considered to be relevant management solutions for the wetland (Fig. V.5-1). Finally, more detail on the individual scores per indicators can be found in section VI.5.

## V.6  Concluding remarks

A qualitative analysis was conducted on the Abras de Mantequilla wetland based on expert elicitation. This information source was needed since there was very scarce or even no quantitative information, contrary to the water quantity and quality indicators. The Lickert scale was a key tool to translate subjective perceptions into numerical values. For each indicator, a value function was provided, either through sources, namely the WET-Health methodology or mathematical expressions such as the quadratic function.

Based on their area of expertise, the panel scored the performance of each proposed management solution facing the criterion the indicator aimed to. The scores tended, in most of the cases, to be higher as the complexity of the MS increased. These were the MS5 and MS4, resp. Worth noting is to highlight the similarity of these preferences with the results obtain after the modeling framework (Chapter IV), where MS4, followed very closely by MS5, obtained the highest score as the best MS. Finally, given these two important components, viz. both quantitative and qualitative indicators, the final step was to incorporate the stakeholders' preferences and integrate all sources of information into a system to support further management decisions (Chapter VI).

# VI Decision Support Framework for the AdM wetland-river system*

## VI.1 Theoretical considerations

### VI.1.1 General concepts

A Decision Support System (DSS) can be defined as the computational tool that facilitates several issues to a decision maker. Amongst them there are, the problem understanding, the integration of different components, and the interactive selection of the *best* choice. Everything in order to solve a complex, often multidisciplinary challenge. Therefore, that tool demands many knowledge information sources: data, models, experience and others (Rizzoli and Young, 1997). In a simpler way according to some literature, DSS is a computer-based system that assist in the decision making process (Finlay, 1994; Sprague and Watson, 1993) to solve unstructured situations (Gorry and Scott Morton, 1971).

The main aim of a successful DSS is to reduce the complexity inherent into any decision making process due to the large number of involved variables and interactions. In this regard, a DSS has the capability to convert data into knowledge information (Miller et al., 2004). Simultaneously, potential trade-offs between several targets or elements (e.g. stakeholders) should be tackled by these systems (Witlox, 2005). And it is important to acknowledge that every member of this *community* can provide a sort of knowledge: either traditional, empirical, experience-based, science-based or just *common sense* (Jonoski, 2002).

Several kinds of DSS are available depending on their target domain or the problems to tackle. Amongst the most typical there are spatial decision support systems (SDSS), environmental DSS (EDSS), planning support systems (PSS), multi-criteria DSS, etc.

* Parts of this chapter are based on: Arias-Hidalgo, M., Villa-Cox, G., van Griensven, A., Solórzano, G., Villa-Cox, R., Mynett, A.E., Debels, P. (2012). "A decision framework for wetland management in a river basin context: the "Abras de Mantequilla" case study in the Guayas River Basin, Ecuador", <u>Journal of Environmental Science & Policy, Sp. Ed.</u> (accepted for publication), doi:10.1016/j.envsci.2012.10.009.

Depending on the information source or carrier they can be classified as: communication-driven, data-driven, knowledge-driven, model-driven DSS and more recently web-based DSS or a combination of several sources (Turban, 2007).

DSS are more suitable to be applied on the so-called *unstructured or semi-structured* problems (Blanning and King, 1993; Keen and Scott Morton, 1978; Sprague and Watson, 1993). The more data available or the more preparedness, skills or knowledge the decision makers have, the more structured the problem is and viceversa (Witlox, 2005). Model-based analyses are useful in case of full structured systems whereas knowledge-based DSS are in the unstructured ones. For combined situations (such as the present study) where information comes from both quantitative and qualitative assessments, a mixed approach might be more suitable.

Four main causes contribute to the unstructuredness of the situation. First, the lack of appropriate criteria, pros & contras about each proposed solution. Secondly, non-suitable initial weight assignation to each criterion that leads to unfair or non-existent negotiation or tradeoff amongst the existent criteria or objectives. Third, wrong constraints or restrictions, whose consequences are unknown in most of the cases; and finally, not sufficient data that may have allowed eliminating repeated solutions or rise hidden ones (Armstrong et al., 1990).

Of course, the target when building a DSS is to reduce unstructuredness, i.e. provide clearance to the decision maker. Some authors propose two general ways in this regard. First, to base DSS on developing a set of end-mean relationships that depend on the scale (Reitsma, 1990) and as long as the target is reached through a low number of solutions (i.e. the problem is not too abstract). A second alternative is to divide the unstructured problem in a series of connected structured and solution-prone stages, where more transparency and speed can be achieved than in the former variant (Witlox, 2005).

## VI.1.2  Multi-criteria Analysis and decision space

The core of a DSS is to be capable to evaluate several possible options. For this, a DSS uses the so-called Multi-criteria Decision Analysis or just Multi-criteria Analysis (MCA). As its name indicates, MCA attempts to find a *best compromise solution* that sufficiently satisfies all involved goals (Simon, 1981). This situation is typical in the water sector as decision makers (DM) cope with several interests coming from different sorts of stakeholders. Knowing that each stakeholder interest or potential action have positive or negative consequences, the DM's target is to attain a certain interest without damaging or affecting negatively others and so forth. To that end, the involved actors need to assess the advantages and disadvantages of each alternative through a set of decision rules (Sen and Yang, 1998).

Suppose a finite set of options: $O = \{o_1, o_2, \ldots\ldots, o_n\}$ and a set of targets, goals or criteria, $G = \{g_1, g_2, \ldots\ldots, g_k\}$. $G_j(o_i)$ represents the evaluation of criterion $g_j$. Every $G(o_i)$ needs to be maximized (e.g. for profits) or minimized (e.g. for errors or losses). As a result, a decision matrix (to be discussed in section VI.5) can be defined as follows (Bogardi, 1994; Genova et al., 2004):

$$A = \begin{bmatrix} a_{11} & a_{12}\ldots\ldots.a_{1k} \\ a_{21} & a_{22}\ldots\ldots.a_{2k} \\ . & . \ldots\ldots . \\ . & . \ldots\ldots . \\ a_{n1} & a_{n2}\ldots\ldots.a_{nk} \end{bmatrix}$$

Eq. VI-1

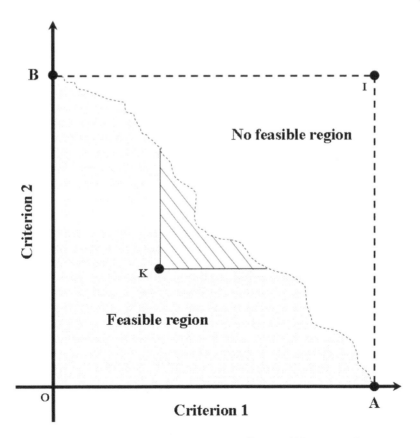

Fig. VI.1-1 The Pareto front and the decision space (Sen and Yang, 1998).

In order to understand how two variables (with their own interests) interact within a decision making process, two concepts must be expressed. The Decision Space is defined as the domain where the variables values are located; whereas the criterion outcome space (a.k.a. payoff space) is the domain formed by the consequences associated to these variables (Malczewski, 1999; Tecle and Duckstein, 1994). In Fig. VI.1-1, a decision space is shown where points A & B represent the maxima satisfactions for criteria 1 & 2, respectively. I is the *ideal* place where both interest should meet theoretically.   The hatched area marks a more realistic region when these two demands can co-exist each other, being any point dominant or *better* solution than the pair K.   The point O, in contrast and although part of the *feasible* field, it is not advisable because none of the criteria are met at all.   For the points on the dot line (or frontier between the feasible and no-feasible solutions) there is no alternative that dominates them.   This non-dominated set (Malczewski, 1999) is known as the *Pareto* front.  When one of the manifold interests improves without compromising or damaging others then such a move is called a Pareto improvement.   The ideal outcome would be to find the most suitable point on this front.   Figures Fig.   VI.1-2 and Fig.   VI.1-3 show examples of decision and criterion outcome spaces, respectively.

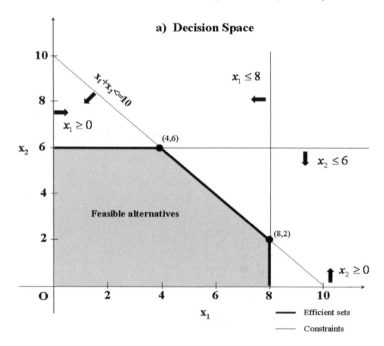

Fig.   VI.1-2 Constraints, efficient sets in the decision space in the decision space (Malczewski, 1999).

Rarely there is a perfect or unique solution when many criteria are simultaneously aimed.   This happens because physical, economical, or social conditions / restrictions do not allow it or make theoretical solutions impractical or no-feasible.

These constraints reduce the decision space for feasible solutions as seen in Fig. VI.1-2.

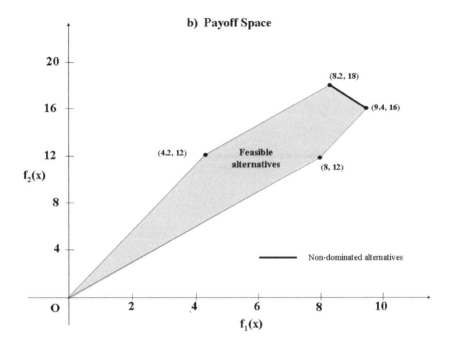

Fig. VI.1-3 Non-dominated alternatives in the criterion outcome space (Malczewski, 1999; Tecle and Duckstein, 1994)

Sen & Yang (1998) classifies two kinds of MCA methods. Firstly, the Multiple Attributes Decision Making (MADM) which selects the best alternative from a set of options based on their most important attributes. The criteria (option's performance) can be fairly quantified using normalized scales via value functions (as in Chapter IV and V for the quantitative and qualitative indicators). Thus, very high/low quantified attributes will not have exaggerated or too minimal contributions to the final score. Since MCA is often a multidimensional problem, the purpose is to synthesize it into a ranking of alternatives based on individual criteria later on aggregated to overall preference. To perform this task, some decision rules are available (i) Simple Additive Weighting (SAW); (ii) Order Weighting Average (OWA); (iii) TOPSIS (Technique for Order Preference by Similarity to Ideal Solution) (Hwang and Yoon, 1981; Yoon and Hwang, 1995); and (iv) ELECTRE (Nachtnebel, 1994) and others. This study focuses in this particular MCA category. The second MCA method is the Multiple Objective Decision Making, MODM, which selects the most appropriate choice based on the prioritized goals and taking in consideration existent constraints (Fig. VI.1-4). In general, it can be stated that objectives are attributes (e.g. cost / profits) with specifications. No list of solutions is provided, just requirements. These requirements, for instance, minimizing costs or maximizing profits can be numerically solved by means of linear or non-linear programming techniques.

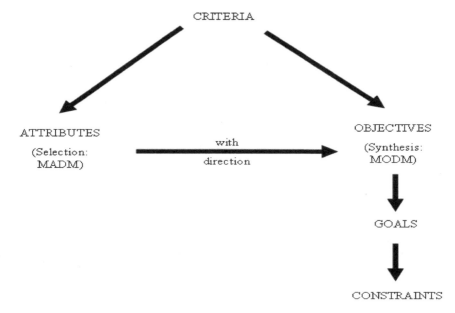

Fig. VI.1-4 Multiple Criteria Decision Making methods (Sen and Yang, 1998).

## VI.2  Decision rules

### VI.2.1  Simple Additive Weighting (SAW)

The Simple Additive Weighting decision rule (SAW) is a very popular technique due to its simplicity. It uses a composite summation of subjective weights (criterion weights) multiplied by their decision scores, in the following way:

$$\phi_s(o_i) = \sum_{j=1}^{n} w_j * d_{ij}$$

Eq. VI-2

where:

$\phi_s(o_i)$ : SAW for each option i;

$w_j$: Criterion weight of each indicator, usually derived from a stakeholder analysis;
$d_{ij}$: Normalized decision scores (from the evaluation matrix).

## VI.2.2  The TOPSIS method

TOPSIS (Hwang and Yoon, 1981; Yoon and Hwang, 1995) is a MADM technique which selects the closest options to a certain *ideal* point and the farthest to a *nadir* or worst location. Relatively simple, the tool requires two input data: a decision matrix (Eq. VI-1) and weights. The latter mathematically express the preferences (e.g. from stakeholders). To compute how far a point is from the target locations, TOPSIS uses the Euclidean distance and assumes that the utility of each attribute decreases or increases proportionally. However, a potential drawback may reside on the subjectivity (thus high uncertainty) of the selected weights.

The elements from the decision matrix are normalized as follows (Sen and Yang, 1998):

$$z_{ij} = \frac{a_{ij}}{\sqrt{\sum_{i=1}^{n} a_{ij}^2}}, \qquad i = 1,...n; \quad j = 1,...k$$

Eq. VI-3

For every $i^{th}$ option, each $w_j$ ($j^{th}$ weight) is combined with the normalized elements of the decision matrix to produce the weighted normalized decision matrix, whose elements are like this:

$$x_{ij} = w_j z_{ij}, \qquad i = 1,...n; \quad j = 1,...k$$

Eq. VI-4

Now, assuming that the ideal point is the maximum *benefit* (B) and the nadir implies the maximum *cost* (C), the set of target locations per attribute for each $i^{th}$ option is defined, respectively, as:

a) The ideal set:

$$a^* = \left\{ (\max_i \ x_{ij} \,\middle|\, j \in B), \ (\min_i \ x_{ij} \,\middle|\, j \in C) \,\middle|\, i = 1,.....n \right\}$$

$$a^* = \left\{ x_1^*, x_2^*, .....x_k^* \right\}$$

Eq. VI-5

b) The non-ideal set:

$$a^- = \left\{ (\max_i \ x_{ij} \Big| j \in C), \ (\min_i \ x_{ij} \Big| j \in B) \Big| \ i = 1,.....n \right\}$$

$$a^- = \left\{ x_1^-, x_2^-, .....x_k^- \right\}$$

Eq. VI-6

From each element $x_{ij}$, the Euclidean distances to the $a^*$ & $a^-$ elements are:

$$D_i^* = \sqrt{\sum_{j=1}^{k} (x_{ij} - x_j^*)^2} \quad i = 1,....n$$

Eq. VI-7

$$D_i^- = \sqrt{\sum_{j=1}^{k} (x_{ij} - x_j^-)^2} \quad i = 1,....n$$

Eq. VI-8

Furthermore, for every $i^{th}$ option, the relative closeness (between 0 and 1) is computed in the following way:

$$C_i^* = \frac{D_i^-}{(D_i^- + D_i^*)}$$

Eq. VI-9

Finally, the selection of alternative depends on the closeness values. The higher the closeness, the higher the rank and viceversa.

## VI.3  DSS in catchment-wetland systems

As it has been mentioned before, the decision support system plays a key role in providing the necessary information to the decision maker in order to choose the best management strategy. Water systems are not the exception in this regard, since every water environment (either urban, fluvial, wetland, coastal or irrigation management system) requires an appropriate decision making process. Unfortunately, most of existent DSS in the water sector are still incomplete in their conception. The information they collect is mostly devoted to the use of mathematical models and characterization. More attention should then be given to propose a solution or a set of solutions or to perform a MCA analysis to support a decision process. Systems that analyze management options too close to the BAU status or far from realistic alternatives (that could be applicable in practice) should be enhanced. Actually, to keep the *status quo* is no longer considered a valid alternative or a solution (Bostrom and Suzina, 2009).

Examples of applications of decision support tools have been conceived and implemented for river catchments, but often with scarce considerations for wetlands (Barrow, 1998; La Jeunesse et al., 2003; van Ast, 2000; Welp, 2001; Williams, 2001). Conversely, some interesting DSS applications have been addressed just for wetlands but outside a rural catchment context (Walters and Shrubsole 2003; Kirk, Wise et al. 2004; Cabrera 2008). As a result of this target mismatch, different management strategies and planned policies and consequent inapplicability on these twinned systems could have been the most likely outcome in previous studies.

## VI.4 NetSyMoD and MDSS5

The NetSyMoD methodological network (social NETwork analysis creative SYstem MOdelling and Decision support approach) was an initiative of several European institutions. Its main targets were the diffusion of information and decision support in natural resources planning and management (Giupponi et al., 2006). As a part of this system, the mDSS software package was developed (Giupponi et al., 2010) as a further stage of the early MULINO (MULti-sectoral Integrated and Operational DSS for Sustainable Use of Water Resources at the Catchment Scale) computational program. In general NetSyMoD attempts to support decision makers to select the most suitable options, taking in consideration several approaches. The tool not only permits stakeholder involvement in the decision making process but also incorporates social and economic issues to the traditional environmental viewpoint in each of the alternatives that are envisaged and proposed. In fact, the inclusion of social and economic issues had been already recommended for a better water resources management (Balloffet and Quinn, 1997; Lovejoy et al., 1997).

Throughout the NetSyMoD tool (Fig. VI.4-1), six stages are defined (i) Stakeholder Analysis, or identification of the main actors, their roles and interactions in the case study; (ii) the Problem Analysis, where the DPSIR chains are determined including the definition of indicators; (iii) the Creative System Modeling, where a modeling framework is adopted based on the data and information availability and the involved indicators; (iv) the DSS design, where all sources of information (either quantitative and qualitative) are normalized and integrated; (v) the options analysis, where the management alternatives (under the potential scenarios) are formally proposed, discussed and evaluated, resulting in a final ranking of solutions, based on the preferences and decisions by the stakeholders; and finally (vi) the actions &

monitoring phase, which means the process is iterative meanwhile the system is being implemented, adapted, receives feedback and ultimately is disseminated.

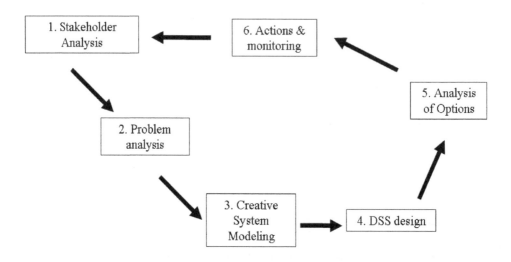

Fig.  VI.4-1 NetSyMoD methodology flowchart (Giupponi et al., 2010)

The mDSS program (v.5) was specifically designed for the DSS design stage. Beside the easy and straightforward implementation, according to Giupponi, Cojocaru et al. (2010), this conceptual tool has some other advantages (i) it improves the user's understanding of the nature of various sorts of stakeholders (e.g. governance, local) and their interactions; (ii) with the help of pre-conceived management options, mDSS5 enables room for better decisions; (iii) encourages the public participation (when consulted); (iv) enlarges the cooperation between different and within each group of stakeholders; and (v) facilitates the comprehension of the stakeholder preferences.

Sequentially, the mDSS has three main phases (i) the conceptual; (ii) the design; and (iii) the selection (Fig.  VI.4-2).  In the conceptual phase, the main involved issues and problem definition take place.  A local network analysis (stakeholders) and the DPSIR chains identifying drivers, states, impacts and responses.  The latter are the basis for the future management options.  Exogenous Drivers (ED), such as climate changes or the Baba project in the AdM case, are incorporated in this stage as a part of a more elaborate scenario-based DPSIR chain.

The design phase begins with the identified responses.  To this purpose, each indicator has to be characterized via quantitative (models) or qualitative (data, expert knowledge) for the current status and for each possible response.  This leads to the definition of the final proposed management options.  Hence, the management options and the involved criteria of each stakeholder are organized in the Analysis Matrix (AM, already mentioned in section VI.1.2). Since the scales are dissimilar amongst the indicators, the resultant values are normalized resulting in the Evaluation Matrix (EM). The normalization is carried out via value functions which depend on expert knowledge, existent literature or the preferences of decision makers and stakeholders.

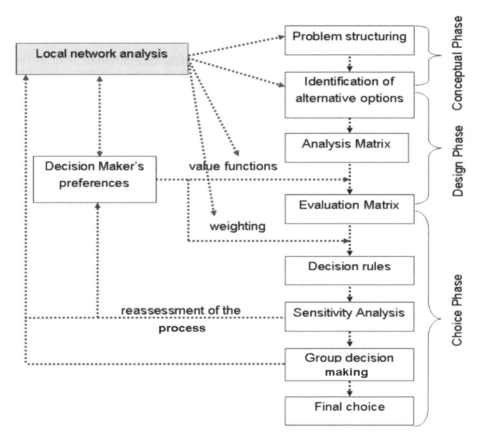

Fig. VI.4-2 mDSS work flowchart (Giupponi et al., 2010).

Finally, the *choice phase* starts with the application of weights to the already computed evaluation matrix. These weights are derived from the stakeholder analysis as well as the decision maker's choices. Thus, the MCA analysis takes place in order to choose the best alternative, using any of the available decision rule methodologies, mentioned earlier in this chapter. The process is iterative as some indicators and options can be refined (sensitivity analysis). Ultimately, the individual preferences are combined leading to a group decision making process and to the final selection of the *best* solution.

## VI.5 mDSS application on the AdM case study

As an integrative stage beyond the modeling framework and the expert elicitation, the mDSS tool was applied to the AdM case study. The DPSIR chains, identified in chapter II.3, were the basis for the preliminary analysis or conceptual phase (Fig. VI.5-1). The main exogenous drivers, the major infrastructure works / the climatic variations (*clich_infr*) and the landuse degradation (*pop_tr*) were part of the BAU scenario, to which every Management Solution (MS) was compared against.

Each component had associated indicators depending on the nature. For instance, the driver crop production (*crop_prod*) was derived from the population trend (land degradation for food production) since obviously more people in the wetland demand more food, leading to a pressure (degradation - *degrad*) which may have affected one of the states (e.g. water quality – *wtr qlt*). As an impact, eutrophication (*eutroph*) might have been an effect on the system; therefore a response was also expected from it, in the figure of stakeholder capacity to take a decision (*stak_capac*). Last but not least, the proposed management solutions were applied to any of the distinct stages (D, P, S or I), depending on the indicator and its respective management axe (to be discussed shortly).

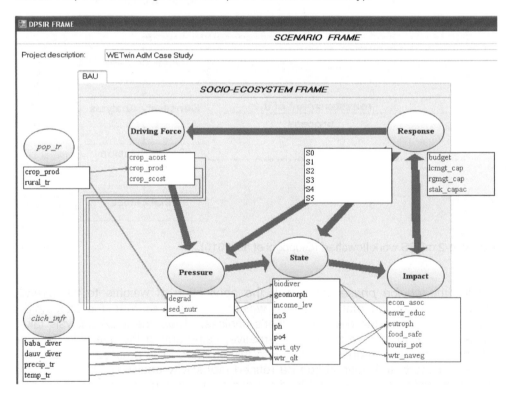

Fig. VI.5-1 DPSIR chains as input data for mDSS5.

As a result of the quantitative and qualitative analysis of the indicators, an analysis matrix was built (Table VI.5-1). In this matrix, the rows were the criteria or indicators in their various categories: socio-economic, institutional, biological and quantitative. On the other hand, the columns were the proposed management solutions (in total 19 x 6). The values on that matrix indicated the performance of each management solution with respect to the indicator. For instance, in *food safety for local people* (food_safe), options #3 and #5, and in lesser degree #4, had higher scores than the rest, according to the criteria of scholars/experts. Nonetheless, as stressed before, the analysis matrix was not yet suitable for use, since the scales were not the same for all indicators (raw performance). Therefore, all scores were normalized using the proposed value functions (relative performance). This normalization was the basis for the evaluation matrix (Table VI.5-2).

Moreover, the relevance of every ecosystem service had to be quantified. Such weights were the result of two workshops with local stakeholders (66 people, mostly farmers, fishermen, and some NGOs) as well as with decision makers such as the Technical Secretariat of the AdM Commonwealth of Municipalities. In those meetings, the interviewed people were given a list of management axes or categories which were related to the ecosystem services that the wetland provides according to the findings after following the WET-Ecosystem evaluation (Kotze et al., 2008) (Chapter II.2). By means of this mechanism, the stakeholders (especially the local ones) could properly relate their own view of the environment with formal ecosystem terminology. Thus, they freely chose or gave their preferences and a tradeoff could take place.

The authorities gave their opinion about the ecological services (Fig. VI.5-2). For them, water quality control was a key factor, driven most likely due to a previous controversial decision of a former major of Vinces town to dispose the urban garbage in one of the corners of AdM (Noroña, 2009). In second place was the availability of drinking water (related to water quantity). Thus, they acknowledged the potential of the wetland to store water especially during the dry season. Biodiversity occupied the third place. Perhaps the reason could be the eco-touristic potential these municipalities foresee for the wetland in a nearby future. Finally, cultural and institutional significance was also relatively important for the majors. This is not surprising since the Commonwealth of Municipalities was precisely created to strengthen the links between neighbouring local governments taking advantage of similar needs, problems, socio-economic activities and environmental issues. On the other hand, services such as carbon storage or extraction of natural resources (neither agricultural nor livestock activities) are, according to them, not that relevant. In the first case, this is likely due to lack of understanding or ignorance. In the latter, it might be related to their concern for biodiversity, since they considered that wood exploitation or large-scale fish industry, for instance, might be more harmful than beneficial to the wetland ecological status. Unfortunately, the future views on ecosystem services by the municipalities and ministries could not be obtained because in most of the cases these politicians were not interested in what their successors might do afterwards.

Table VI.5-1   Analysis Matrix. SE: Socio-economic indicators; B: Biological indicators; I: Institutional indicators; Q: Quantitative indicators.

| Indicator code | Indicator name | Description | MS 0 | MS 1 | MS 2 | MS 3 | MS 4 | MS 5 |
|---|---|---|---|---|---|---|---|---|
| SE1 | crop_acost | Crop maintenance cost per Ha. | 3.0 | 3.2 | 3.0 | 3.0 | 2.8 | 2.0 |
| SE2 | crop_scost | Crop sowing/investment cost per Ha. | 3.0 | 3.0 | 2.6 | 2.8 | 2.0 | 1.6 |
| SE3 | food_safe | Food safety for local populace | 3.0 | 2.8 | 2.9 | 3.5 | 3.4 | 3.5 |
| SE4 | crop_prod | Crop productivity per Ha. | 3.0 | 3.3 | 3.7 | 3.7 | 3.7 | 3.7 |
| SE5 | sed_nutr | Sediment and nutrient contribution of landuse | 3.0 | 3.4 | 4.2 | 4.4 | 4.4 | 4.8 |
| SE6 | income_lev | Local populace income level | 3.0 | 3.4 | 3.7 | 4.0 | 4.0 | 4.4 |
| SE7 | econ_asoc | Economic association potential of local stakeholders | 3.0 | 3.3 | 3.5 | 3.5 | 3.0 | 3.0 |
| SE8 | wtr_navig | Water navigability of the wetland | 3.0 | 4.2 | 4.2 | 4.6 | 4.6 | 4.6 |
| SE9 | touris_pot | Touristic potential of AdM | 3.0 | 3.6 | 3.4 | 3.5 | 4.1 | 4.4 |
| SE10 | envir_educ | Level of environmental conscience and education of local populace | 3.0 | 3.5 | 3.9 | 4.1 | 4.8 | 4.8 |
| B1 | eutroph | Eutrophication indicator (estimated by the presence of water hyacinths) | 3.0 | 3.8 | 4.0 | 4.3 | 3.8 | 3.8 |
| B2 | biodiver | Biodiversity | 3.0 | 3.4 | 3.4 | 3.4 | 4.6 | 4.8 |
| I1 | stak_capac | Local stakeholder capacity | 2.3 | 2.5 | 2.8 | 2.8 | 3.3 | 3.3 |
| I2 | lcmgt_cap | Local Management structure capacity (municipalities, NGOs, etc). | 3.0 | 2.8 | 3.0 | 2.8 | 3.3 | 3.0 |
| I3 | rgmgt_cap | Regional/national management structure capacity to coordinate action and support initiatives | 2.3 | 2.7 | 2.7 | 3.0 | 3.3 | 3.7 |
| I4 | budget | Budget adequacy to undertake Management Solutions. | 1.3 | 2.0 | 2.0 | 2.3 | 2.7 | 2.7 |
| Q1 | wtr_qlt | Water quality (Index) | 91.0 | 92.0 | 95.0 | 95.0 | 95.0 | 95.0 |
| Q2 | wtr_qty | Water quantity (Hm3) | 29.2 | 37.7 | 38.3 | 38.1 | 38.9 | 38.7 |
| Q3 | degrad | Land degradation | 8.4 | 8.4 | 8.1 | 7.9 | 6.9 | 6.7 |

Table VI.5-2 Evaluation matrix.

| indicator | MS0 | MS1 | MS2 | MS3 | MS4 | MS5 |
|-----------|-----|-----|-----|-----|-----|-----|
| crop_acost | 0.57 | 0.59 | 0.57 | 0.57 | 0.54 | 0.44 |
| crop_scost | 0.57 | 0.57 | 0.52 | 0.54 | 0.44 | 0.40 |
| food_safe | 0.57 | 0.54 | 0.55 | 0.62 | 0.60 | 0.62 |
| crop_prod | 0.57 | 0.60 | 0.63 | 0.63 | 0.63 | 0.63 |
| sed_nutr | 0.57 | 0.61 | 0.73 | 0.80 | 0.80 | 0.93 |
| income_lev | 0.57 | 0.61 | 0.63 | 0.66 | 0.66 | 0.81 |
| econ_asoc | 0.57 | 0.59 | 0.62 | 0.62 | 0.57 | 0.57 |
| wtr_navig | 0.57 | 0.73 | 0.73 | 0.86 | 0.86 | 0.86 |
| touris_pot | 0.57 | 0.63 | 0.60 | 0.62 | 0.70 | 0.79 |
| envir_educ | 0.57 | 0.62 | 0.65 | 0.70 | 0.92 | 0.92 |
| eutroph | 0.57 | 0.64 | 0.66 | 0.75 | 0.64 | 0.64 |
| biodiver | 0.57 | 0.61 | 0.61 | 0.61 | 0.86 | 0.93 |
| stak_capac | 0.47 | 0.50 | 0.54 | 0.54 | 0.59 | 0.59 |
| lcmgt_cap | 0.57 | 0.54 | 0.57 | 0.54 | 0.59 | 0.57 |
| rgmgt_cap | 0.48 | 0.53 | 0.53 | 0.57 | 0.60 | 0.63 |
| budget | 0.37 | 0.44 | 0.44 | 0.48 | 0.53 | 0.53 |
| wtr_qlt | 0.82 | 0.84 | 0.90 | 0.90 | 0.90 | 0.90 |
| wrt_qty | 0.76 | 0.86 | 0.87 | 0.86 | 0.87 | 0.87 |
| degrad | 0.16 | 0.16 | 0.19 | 0.21 | 0.31 | 0.33 |

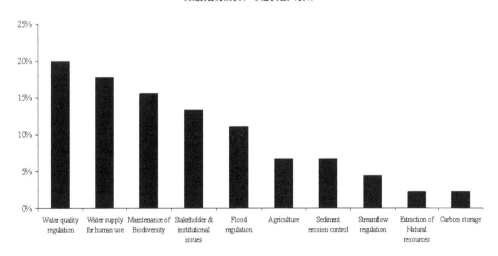

Fig. VI.5-2 Current relevance of Ecological Services in AdM according to government stakeholders.

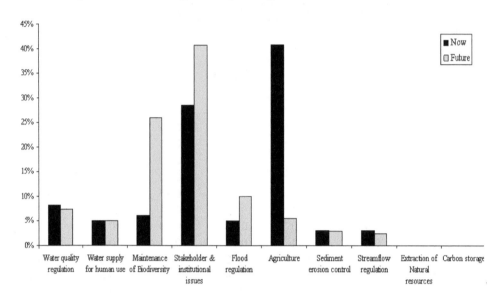

Fig. VI.5-3 Current and future relevance of Ecological Services in AdM according to local stakeholders.

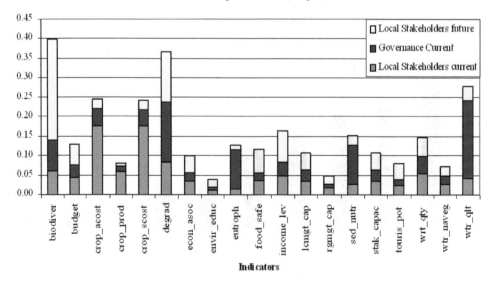

Fig. VI.5-4 Weights of each indicator, for each group of stakeholders.

When the local stakeholders were consulted, their views were sometimes different than the ones from the authorities (Fig. VI.5-3). Besides, the local people did show both their present and potential views in the future. Nowadays, they consider agricultural activities as the main ecosystem service the wetland can provide. This was expected since most of these dwellers are farmers. Issues such as institutional cohesion and socio-economic activities such as eco-tourism and maintenance of biodiversity received a lot of attention as well, especially for the nearby future. In this regard, they see themselves as eco-guides. Dwellers are also very concerned about the current status of wild life (e.g. monkeys and local fish species) with respect to invasive and possibly harmful species such as tilapia. Regarding other services such as water supply and flood regulation (related to the water quantity indicator), they did not show much interest probably due to their sufficient and relatively clean water supply from wells. Similarly, other wetland roles, such as streamflow regulation, flood control and extraction do not seem so important, maybe also, as in the case of the government bodies, due to the lack of knowledge on the matter.

All these partial weights were just referred to the management axes (categories) but not yet to every indicator. To reach those values, it was necessary to establish which indicators belonged to a certain axe and within that axe which was the weight of each indicator. For instance, the *agriculture* category comprised the following indicators: *crop_acost* (Crop maintenance cost per Ha.), *crop_scost* (Crop sowing/investment cost per Ha.) and *crop_prod* (Crop productivity per Ha.). The relative weights that each had with respect to the category depended on a certain ranking. In this example, it was considered that both investment and maintenance costs (*crop_scost* and *crop_acost*) were more influencing than *crop_prod* but equally important between each other. Therefore the ranking was 3, 3 & 1, and the partial weights inside *agriculture* were, respectively, 3/7, 3/7 and 1/7. Finally, the partial weights were multiplied by the axes' weight giving thus the overall weight per indicator. Fig. VI.5-4 shows the indicator weight values per stakeholder group and in the case of the local ones, current and future views as well.

## VI.6 Results & Discussion

Combining the found weights after the workshops with the evaluation matrix, a ranking of management solutions was obtained. First of all, it seems that the current priorities between the local governments and local stakeholders were not entirely compatible. The latter preferred MS4 which was in line with what was found by the WET-Ecosystem evaluation, especially linked to cultivated foods (crop substitution & agricultural practices) and maintenance of biodiversity (reforestation). In this respect, the crop maintenance and investment costs/Ha held the largest weights (in average amongst the MS) with 16 & 15% respectively (which actually meant more reduction on costs according to the experts in the workshops), followed by water quantity (8%, related to crop irrigation). Although biodiversity (ecological corridors) introduced an important contribution to the preferences (MS5 is the second best in the ranking), MS4, with its moderate crop substitutions and less crop costs (compared with MS5), still remained as the chosen management solution by the local people (Fig. VI.6-1a).

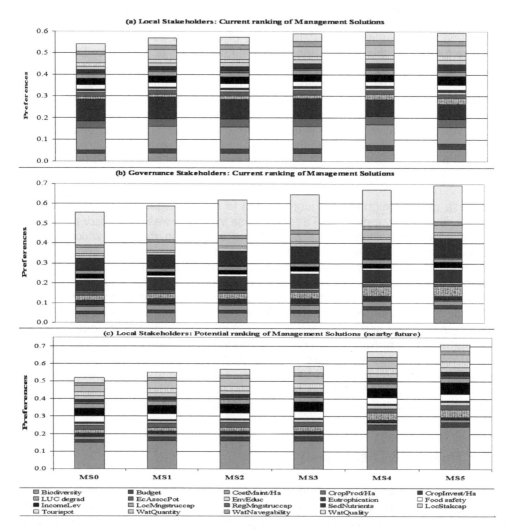

Fig. VI.6-1 Above, (a): Local stakeholders current MS preferences; Middle, (b): Government stakeholders MS preferences; Below, (c): Local stakeholders potential MS preferences.

On the other hand, it was observed that governance stakeholders (e.g. AdM municipalities) gave more importance to water quality (related to sanitation) issues than other domains (27% of the total, in average for each MS). This is consistent with other resultant weights such as nutrient/sediment retention (12%) and reduction of eutrophication (10%). But they also regarded biodiversity and lower landuse cover degradation indices as one source of higher income levels and development of touristic potential. In effect, the increment in weights is also due to the aforementioned fields when we moved to more elaborated management solutions (3, 4 & 5). Therefore, the outcome choice was the MS5 (i.e. intensive landuse substitution and use of ecological corridors) (Fig. VI.6-1b).

Notwithstanding these preliminary dissimilar preferences, future expectations of local stakeholders show that if reinforcement and management are sustained, the two current views may converge to similar interests. It was found that biodiversity in the nearby future is of major concern of local farmers and fishermen (in average 30% of the total weight). Other major factors were the income level and food safety, which may suggest a shift in their daily activities. The touristic potential also contributes to the top ranking of the MS5 option, since several stakeholders consider ecotourism as an important source of revenues in the upcoming years (Fig. VI.6-1c).

Finally, in each of the three instances and for the socio-economic indicators, independently of which of the four proposed value functions was used (section V.3), there were no significant changes in the indicator weights at each MS and thus, no variations in the final ranking.


## VI.7 Concluding remarks

A simple decision support methodology has been applied to the Abras de Mantequilla case study in Ecuador. The system was mainly based on quantitatively analysis (a modeling framework) and qualitatively analysis (expert elicitation). The different sort of outcomes from the analyses, referred to each of the proposed management solutions were harmonized in an evaluation matrix by means of selected and appropriate value functions. A set of weights, one for each indicator, holding the influence of each on the final ranking of management solution was calculated after workshops with the involved stakeholders: local inhabitants (usually farmers and fishermen) and governmental (mostly municipalities and ministries). Management axes are categories that comprise several indicators. Using the influence that each category exerted on the global set and the weight that each indicator held on its respective axe (category), it was possible to determine the overall percentages of each indicator.

When a ranking of management solutions was obtained, the most elaborated alternative (MS5) was preferred amongst government stakeholders (municipalities). On the other hand, the local stakeholders (wetland inhabitants), when consulted about their present and future interests, were of the opinion that small-scale crop substitution and some reforestation would be the best way to restore (or to start restoring) the wetland. At the same time, they also regard AdM as an ecotourism hub for the Los Rios Province in a near future, as well as consider more radical changes in landuse cover (e.g. cocoa instead of corn); thus MS5 may also be their choice later on.   In general, the DSS tool developed by the WETWin project remains open for more and better data and enhancement, for instance, to clarify even more the evaluation of some management axes/indicators (e.g flood control).

These findings open opportunities for mutual cooperation directed to facilitate the territorial ordering process that is currently taking place in the AdM municipalities by command of the central government in Ecuador.   By allowing space for future negotiations amongst the actors, this methodology is also amenable to continuous improvement towards a better tradeoff and wetland/river catchment management. Such actions may comprise a deeper feedback from more local stakeholders, e.g. NGOs and future interests of government stakeholders, for instance SENAGUA (National Ministry of Water), as well as enhancing the communication between scientists and actors in order to refine their preferences.   After all, to communicate in a sound way and to understand and speak each other in the same *language* are still challenging issues in the water sector, as it has recently been emphasized in several discussions during the 6th World Water Forum in Marseille, France (2012).

$$\boxed{\textit{Chapter Seven}}$$

# VII    Conclusions and recommendations

## VII.1 Conclusions

Nowadays, there is a growing recognition on the need to integrate wetlands as important elements within the context of a river basin management. This integration should not be accomplished only at the modeling level but also at the decision making stage. A case study within a European-sponsored project has been selected as a pilot study to conduct a methodology. It is the Abras de Mantequilla, a Ramsar site, located in the middle part of the Guayas River Basin (34000 $Km^2$) in Ecuador. Since a specific feature of this location was poor data availability, this imposed a default limitation on the attainable complexity of the modeling tools. Not only quantitative (modeling framework) but also qualitative (expert elicitation) ways were employed as complementary approaches to facilitate the decision making process. To this end, a Decision Support System for a tropical wetland – river catchment was developed. The DSS took into account the different objectives of the stakeholders as well as several ecological services (management axes) provided by the wetland.

To start with the methodology, the case study was characterized both at the catchment and local scale. Based on a DPSIR analysis, the main drivers that may exert some sort of pressure or influence on the system were identified. This gave origin to a proposed set of management options to allow the system adaptation to these influencing exogenous agents, namely the Baba Dam construction and the climate changes (together referred as the BaU scenario). The management options ranged from increasing the water retention capacity of the wetland to landuse substitution and ecological recovery. These were combined into different arrays and resulted in management solutions.

Lack of data in the AdM case study, albeit being a difficulty for in-depth research, was also a challenging one. Actually, it provided an opportunity to invoke data mining solutions to cope with the data gaps. One of the variables showing missing data was the river discharge time series. A simple methodology based on the Hodrick-Prescott filter (HPF) was developed for some streamflow measurement stations along the Vinces and the Nuevo River. The HPF approach proved helpful to split the signal between a trend and a noise component throughout the training period and thus allow a pattern characterization. The application went beyond using Fourier series to estimate missing gaps. It was clear that the proposed methodology also had limitations, like random phase shifting, inherent to the $n^{th}$ order Fourier series. Despite this drawback, the patterns for estimating gaps during the wet & dry seasons and the transitions periods were fairly similar to the ones shown by the observed data (comparison of statistical features). When compared to a linear regression between two stations along a river, the latter performed better, as expected. However, given the fact that in several river stations across the Guayas River Basin it was not easy to use spatial or temporal information to estimate patterns for filling in missing data using traditional approaches, the HPF-based methodology from data-mining research proved to have an added value.

The hydrological simulations were helpful to comprehend the seasonal behaviours in the upper Vinces catchment as well as in the upstream part of the Chojampe subbasin. In spite of their simplicity and assumptions, the simulations captured well both the trends and the several peaks occurring during the wet season. Throughout the dry period, the behaviour was acceptable as well, with the exception of some spikes in the transition periods of June and November. In those cases, the model seemed to respond well to a driving rainfall event, whereas the discharge observations exhibited almost no activity. A deeper investigation is required to establish what caused these low-magnitude observations (e.g. measurement technique) or to include a detailed soil infiltration model in the simulation).

New technologies provide manifold options to complement the conventional rainfall spatial data. An example of these is the TRMM (3B42) satellite data. As long as one considers the bias, the type of rainfall and its limitation such as the spatial resolution (not currently applicable to small-scale studies), TRMM data can prove helpful to estimate data gaps. The spatial distribution of the annual rainfall data from TRMM to some extent showed some similarity to the pattern from the ground raingauges. Bias correction factors were calculated and, adopting a simple procedure, were spatially distributed, and thus used to improve the satellite TRMM data, with respect to the calibration measurement spots. Using an empirical, yet effective disaggregation method, it was possible to generate synthetic daily rainfall time series at the satellite spots. These artificial series were incorporated in the existent rainfall-runoff model to complement the ground-based input data and then to assess its performance. The results were quite comparable with those using only gauge information. Consequently, TRMM data can be resourceful in areas where there are no rain gauges such as the Andean foothills in the case of the Guayas River Basin. Finally, there is a potential to estimate gaps in rainfall time series, as the TRMM temporal resolution is usually finer than the available field data. The availability of many products similar to the TRMM 3B42 suggests a high possibility for this task.

Another important data carrier towards the wetland zone was the river routing model. This hydrodynamic simulation confirmed what had been observed in the field but had not yet been scientifically reported, which were the flow interactions between the wetland and the Nuevo River. As expected, the exchange was higher during the wet season when more water went from the river to AdM and around May when it returned to the river. With the exception of the aforementioned isolated events in June and November, the rest of the year the wetland reaches a balance for most of the time.

From a quantitative perspective, the performance of the various management solutions (MS) was gradually increasing with their complexity. The differences between MS1 with respect to the BaU were remarkable compared with the distinction between MS2 (moderate crop substitution) and MS4 (MS2 + ecological corridors), or MS3 (aggressive substitution) and MS5 (MS3 + ecological corridors). This was because the magnitude in variation of the water quantity and quality was much lower when evaluating landuse changes than when storing water. Despite the milder rate of improvement, the distinction between the management solutions was not negligible. When comparing simpler alternatives (e.g. MS1, 2 or 3) to MS5 and MS4, the latter were the most convenient for the wetland management. This happened not only for these involved an increased reservoir but also because MS5 & 4 implied notorious landuse recovery via crop substitution and reforestation.

It should be noted that due to the relatively small size of the wetland, the management options may not lead to strong changes on the Vinces catchment and the Guayas basin. But at a subbasin scale they might have a noticeable effect on the Nuevo River, for instance. Examples of this were the water retention capacity in the AdM wetland and the slight increment in the water quality after the landuse substitution. On the other hand, there was a remarkable influence of river basin scenarios on the wetland management implications. An illustration of this was the expected flow reduction along the Vinces & Nuevo Rivers and consequently at AdM, due to the Baba reservoir project and its partial recovery when an increment in rainfall occurs (due to e.g. climatic variations along the Ecuadorian coastal region)

Knowledge elicitation proved to be another resourceful strategy when data were very scarce. Socio-economic, institutional and ecological issues were evaluated via workshops with local experts and involved stakeholders. From an overall point of view, the consulted experts considered MS5 and MS4 to be the most relevant measures for the managing the wetland and its surroundings. Thus, although being different approaches, the modelling approach and qualitative knowledge elicitation gave similar results and did confirm each other. Finally, the different groups of stakeholders, local inhabitants and municipal authorities had different views on current interests or assessments, but yet related to proposed management solutions (notably MS4 vs. MS5). Nevertheless, these views could well converge in a nearby future, since the future preferences of the local actors were also assessed and indeed approached the present vision of the municipalities for the management of the Abras de Mantequilla and nearby areas. This convergence may facilitate the future development of future management guidelines, in spite of all uncertainties and assumptions made due to data scarcity and the absence of sufficient experiences in the study area.

## VII.2 Recommendations

The ultimate goal of achieving a well managed (sustainable) river basin would best be based on a *holistic* view in which vital elements such as wetlands take a protagonist role, at least for the local and medium catchment scale. The results of the present study have emphasized the necessity to really integrate wetlands into a river basin model before exploring possible management options. This implies including different approaches and disciplines in the basin analysis, notably river hydraulics, catchment hydrology, ecology, land and water development, social considerations, political constraints, etc. This cannot be achieved, in fact cannot even be started in a good manner, if the decision makers are not fully aware of the fundamental necessity of establishing a sound and permanent data collection campaign. That is certainly the case for the Guayas River Basin. There should be increased awareness, habit or even culture of the need for a permanent measuring campaign. Such data collection process should not depend only on whether a big project is to be built or a disaster has occurred. Specific measures could be:

- **To develop a denser network of ground based hydro-meteo stations.** Efforts should be focused specially on the mountains summits and hillslopes, which is currently a deficiency across the basin. As for river stations, every river of high order should have a measurement stations at its catchment outlet and one before or after the junction with another relevant river. Wetlands should also have a decent number of stations inside their domain. A more expanded network will permit the proper use of geo-statistical techniques (e.g. Kriging) to spatially distribute the meteorological variables, thus achieving a better understanding of the spatial dynamics and its initial uncertainties.

- **To acquire a wider set of meteorological variables.** Not only rainfall but also continuous and longer time series on potential evapotranspiration and air temperature should be recorded, at least. At river stations, water stages and flow velocities must be continuously measured and sediment readings should be also incorporated. A full maintenance and re-calibration of the measuring devices and rating curves should be carried out before and after every rainy season, especially when the El Niño phenomenon occurs. Analogous recommendations hold for groundwater measurements (stages) and related important variables such as hydraulic conductivity.

SENAGUA (Water Ministry in Ecuador) is currently planning a pilot program on flood forecasting in the Babahoyo river catchment. To achieve such ambitious target, it is necessary to ensure that good information is available, i.e. go *back to basics*. Without good data, in quantity and quality, projects like this are unlikely to succeed.

Regarding the domain of streamflow series characterization, refinement of the gain rule calculation within the HPF technique applied in this thesis is recommended, e.g. by further research on other types of wavelets. Incorporating more physical knowledge into the mathematical formulation is crucial to improve the understanding and estimate possible data gaps. In addition, alternative ways to decide on seasonal separation and to extend the time series without causing wiggles in the trend during the gap span (thus preventing the current phase-shift) are interesting ideas to be explored.

It is important to keep investigating the use of alternative meteorological data sources. In the case of satellite TRMM data, there is room for more research on ways to improve the disaggregation procedure of time series, from monthly to daily. This methodology can be complemented with the use of spatial information, for instance streamflows at the main outlet and key tributaries, and further small-scale rainfall-runoff simulations. Moreover, a flood simulation (2D) across the wetland area would be a very relevant source of information for further studies. Currently, software modeling systems such as Delft 3D (Deltares, 2011) or LISFLOOD (De Roo et al., 2000; van der Knijff et al., 2010) provide ways to characterize and show the impact of the inundation regime on the wetland and its surroundings. Satellite imagery can be crucial in order to validate this model.

There is still room for further development of the modeling framework, for instance with emphasis on improving the interconnection of models. Since several assumptions and simplifications had to be incorporated along the modeling chain, it is advisable to undertake an uncertainty propagation analysis to provide a suitable probabilistic support to the decision makers. In this regard, reducing assumptions and increasing data density (e.g. groundwater) may entail (i) an enhancement on the analysis of the current quantitative indicators (e.g. water quantity) and their associated management options (e.g. water storage); (ii) that some current qualitative indicators might be objectively quantified (e.g. sediment and nutrient trapping capacity); (iii) the consideration of complementary management options linked to those *new* quantitative indicators; and (iv) a potential refinement of the final ranking of solutions. Nevertheless, this enhanced chain of models should also keep a balance with other sources of information used for extension and verification. Model simulations should not block or obstruct the decision making process but rather be supportive.

Ultimately, for a general improvement of the management process, communication must be highlighted as a key factor. This enhance communication process should obviously involve scientists, decision makers and local actors, to ensure that messages from science, experience, local demands and political views are matching and walking alongside in the same direction. From a regional perspective, in Latin America it is still difficult to devise long-term policies because authorities frequently do not consider about what may happen after their government period. That is the reason why the future views of the government stakeholders could not easily be measured. To tackle this endemic problem, the Ecuadorian government is nowadays seeking for stronger and longer-term policies on agricultural practices, environmental protection and rights of nature (National Assembly, 2008). In addition, multiple perspectives are nowadays taken in account by the national authorities, integrating several ministries and public dependencies and dividing the territory into separate river basins to facilitate the management of natural resources.

# References

Ackerman, C.T., 2009. HEC-GEO-RAS: GIS Tools for support of HEC-RAS using ArcGis, User's Manual, U.S. Army Corps of Engineers, Institute for Water Resources Hydrologic Engineering Center (HEC), Davis, California, USA.

ACOTECNIC, 2010. Estudios de Factibilidad del proyecto DAUVIN: Informe de Hidráulica fluvial (in Spanish). SENAGUA, Guayaquil, Ecuador.

Adler, R.F., Huffman, G.J., Bolvin, D.T., Curtis, S., Nelkin, E.J., 2000. Tropical Rainfall Distributions Determined Using TRMM Combined with Other Satellite and Rain Gauge Information. Journal of Applied Meteorology 39, 2007-2023.

Akaike, H., 1974. A new look at the statistical model identification. IEEE Transactions on Automatic Control 19, 716-723.

Akaike, H., 1980. Likelihood and the Bayes procedure, In: Bernardo, J.M. (Ed.), Bayesian Statistics. University Press, Valencia, pp. 143-166.

Alvarez, M., 2007. Biological Indicators - A tool for assessment of the present state of a river, A pre-impoundment study in the Quevedo River, Ecuador, Environmental Sciences. UNESCO-IHE, Institute for Water Education, Delft, p. 181.

Arias-Hidalgo, M., Villa-Cox, G., van Griensven, A., Solórzano, G., Villa-Cox R., Mynett, A.E., Debels, P., 2012. A decision framework for wetland management in a river basin context: the *Abras de Mantequilla* case study in the Guayas River Basin, Ecuador. Journal of Environmental Science & Policy, Sp. Ed., (accepted for publication), doi:10.1016/j.envsci.2012.1010.1009.

Armstrong, M.P., De, S., Densham, P.J., Lolonis, P., Rushton, G., Tewari, V.K., 1990. A knowledge-based approach for supporting locational decisionmaking. Environment and Planning B: Planning and Design 17, 341-364.

Arriaga, L., 1989. The Daule-Peripa dam project, urban development of Guayaquil and their impact on shrimp mariculture. Coastal Resources Center, University of Rhode Island, Narragansett, R.I., USA.

Bahuguna, A., Nayak, S., Roy, D., 2008. Impact of the tsunami and earthquake of 26th December 2004 on the vital coastal ecosystems of the Andaman and Nicobar Islands assessed using RESOURCESAT AWiFS data. International Journal of Applied Earth Observation and Geoinformation 10, 229-237.

Balloffet, A., Quinn, L., 1997. Special issue on decision-support systems. Journal of Hydrology 199, 207-210.

Barrow, C.J., 1998. River basin development planning and management: A critical review. World Development 26, 171-186.

Bendix, J., Trachte, K., Cermak, J., Rollenbeck, R., NauB, T., 2009. Formation of convective clouds at the foothills of the tropical eastern Andes (South Ecuador). Journal of Applied Meteorology and Climatology 48, 1682-1695.

Bendjoudi, H., Weng, P., Guérin, R., Pastre, J.F., 2002. Riparian wetlands of the middle reach of the Seine river (France): historical development, investigation and present hydrologic functioning. A case study. Journal of Hydrology 263, 131-155.

BirdLife International, 2012. Important Bird Areas factsheet: Abras de Mantequilla (in Spanish). Downloaded from http://www.birdlife.org.

Black, P.E., 1996. Watershed Hydrology, second ed. ed. CRC Press, Boca Raton, FL, USA.

Blackwell, M.S., Maltby, E., Gerritsen, A.L., European, C., 2006. Ecoflood guidelines : how to use floodplains for flood risk reduction. Office for Official Publications of the European Communities, Luxembourg.

Blanning, R.W., King, D.R., 1993. Current research in decision support technology. IEEE Computer Society Press, Los Alamitos, California, USA.

Bloomfield, P., 2000. Fourier analysis of time series : an introduction. John Wiley & Sons, New York, NY, USA.

Bogardi, J.J., 1994. Introduction of systems analysis: Terminology, concepts, objective functions and constraints., In: Bogardi Janos J., Hans-Peter, N. (Eds.), Multicriteria decision analysis in water resources management. UNESCO, International Hydrological Programme, Paris, France.

Borbor-Cordova, M., Boyer, E., McDowell, W., Hall, C., 2006. Nitrogen and phosphorus budgets for a tropical watershed impacted by agricultural land use: Guayas, Ecuador. Biogeochemistry 79.

Bostrom, D., Suzina, A., 2009. 5th World Water Forum tackles Food - Energy Conflicts over Water, In: release, P. (Ed.). World Water Council, Istanbul, Turkey.

Bowman, K.P., 2005. Comparison of TRMM precipitation retrievals with rain gauge data from ocean buoys. Journal of Climate 18, 178-190.

Box, G., Jenkins, G., Reinsel, G., 2008. Time series analysis : forecasting and control. John Wiley & Sons, 746p., Hoboken, N.J., USA.

Brinson, M.M., 1993. A hydrogeomorphic classification for wetlands. Wetlands Research Program, US Army. Corp of Engineers, Waterways Experiment Station, Vicksburg, MS, USA.

Brown, R., McClelland, N., Deininger, R., Tozer, R., 1970. A Water Quality Index - Do we dare?, Proceedings of the National Symposium on Data and Instrumentation for Water Quality Management. Conference of State Sanitary Engineers and Wisconsin University, Madison, Wisconsin, USA, pp. 364-383.

Brunner, G.W., 2010. HEC-RAS River Analysis System, User's Manual Version 4.1. U.S. Army of Engineers (Hydrologic Engineering Center -HEC), Davis, CA.

Buarque, D.C., de Paiva, R., Clarke, R., Mendes, C., 2011. A comparison of Amazon rainfall characteristics derived from TRMM, CMORPH and the Brazilian national rain gauge network. Journal of Geophysical Research 116.

Bullock, A., Acreman, M., 2003. The role of wetlands in the hydrological cycle. Hydrology and Earth System Sciences 7, 358-389.

Burns, A., Bush, R., 2008. Basic marketing research, 2nd ed. Harlow, Prentice Hall, New Jersey, USA.

Cabrera, C., 2008. Framework for selecting MCA methods to support decision making in Wetland Management, Environmental Sciences Department. UNESCO-IHE, Institute for Water Education, Delft.

Carey, J., Zilberman, D., 2002. A Model of Investment under Uncertainty: Modern Irrigation Technology and Emerging Markets in Water. American Journal of Agricultural Economics 84, 171-183.

CEDEGE, 2000. CEDEGE y la planificación de los recursos hídricos en Ecuador (in Spanish), In: Martin, J. (Ed.), 25th anniversary of CEDEGE (Commision of Studies for the development of the Guayas River Basin). Guayaquil, Ecuador.

CEDEGE, 2002. Plan Integral de Gestión Socio-Ambiental de la cuenca del Río Guayas y Península de Santa Elena: informe Edafológico (in Spanish). Commision of Studies for the development of the Guayas River Basin, Guayaquil, Ecuador.

CEDEGE, 2008. Plan Integral de Recursos Hídricos de la Provincia de Manabí: Síntesis y desarrollo cronológico (in Spanish). Commision of Studies for the development of the Guayas River Basin, Guayaquil, Ecuador.

Chaves, H., Alipaz, S., 2007. An Integrated Indicator for Basin Hydrology, Environment, Life, and Policy: The Watershed Sustainability Index. Water Resources Management 21, 883-895.

Chow, V.T., 1959. Open channel hydraulics. McGraw-Hill, New York, NY, USA.

Clement, T.P., 2011. Complexities in hindcasting models--when should we say enough is enough? Ground water 49, 620-629.

Collischonn, B., Collischonn, W., Tucci, C., 2008. Daily hydrological modeling in the Amazon basin using TRMM rainfall estimates. Journal of Hydrology 360, 207-216.

Cornejo, P., 2009. The Abras de Mantequilla wetland, Second Consortium Meeting of Wetwin project. Water and Sustainable Development Centre, CADS-ESPOL, Guayaquil, Ecuador, p. 18.

Coulibaly, P., Burn, D.H., 2004. Wavelet analysis of variability in annual Canadian streamflows (DOI 10.1029/2003WR002667). Water Resources Research 40, W03105.

Cowardin, L.M., Carter, V., Golet, F.C., LaRoe, E.T., 1979. Classification of wetlands and deepwater habitats of the United States. Biological Services, Program, Fish and Wildlife Service, U.S. Dept. of the Interior, Washington, D.C.

Cunge, J.A., 2003. Of data and models. Journal of Hydroinformatics 5, 75-98.

Danthine, J.P., Girardin, M., 1989. Business cycles in Switzerland: A comparative study. European Economic Review 33, 31-50.

De Roo, A.P.J., Wesseling, C.G., van Deursen, W.P.A., 2000. Physically based river basin modelling with a GIS: the LISFLOOD model. Journal of Hydrological Processes 14, 1981-1992.

Debels, P., Zsuffa, I., Namakambo, N., Kone, B., Kaggwa, R., Namaalwa, S., Masiyandima, M., Winkler, P., Hein, T., Pataki, B., Hattermann, F., Liersch, S., Cornejo, P., 2009. Hierarchical framework for data collection, globally and locally available data sets, preliminary characterization of case study wetlands and river basins and development of case study DPSIR chains. VITUKI - SORESMA - ESPOL, Budapest, Hungary, p. 197.

del Río, A., 1999. Agregación temporal y filtro Hodrick-Prescott. Centro de Estudios Monetarios y Financieros, Madrid, Spain.

Deltares, 2011. Delft 3D-Flow User Manual: Simulation of multi-dimensional hydrodynamic flows and transport phenomena, including sediments. Deltares, Delft, the Netherlands, p. 688pp.

Dettinger, M.D., Diaz, H.F., 2000. Global Characteristics of Stream Flow Seasonality and Variability. Journal of Hydrometeorology 1, 289-310.

Dinku, T., Connor, S., Ceccato, P., 2010. Comparison of CMORPH and TRMM-3B42 over mountainous regions of Africa and South America, In: Gebremichael, M., Hossain, F. (Eds.), Satellite Rainfall Applications for Surface Hydrology. Springer Science, New York, USA.

Diskin, M.H., Simon, E., 1977. A procedure for the selection of objective functions for hydrologic simulation models. Journal of Hydrology 34, 129-149.

Duckstein, L., Goicoechea, A., 1994. Value and Utitliy concepts in multiple criteria decision making, In: Bogardi Janos J., Hans-Peter, N. (Eds.), Multicriteria decision analysis in water resources management. UNESCO, International Hydrological Programme, Paris, France.

EC, 2000a. Directive 2000/60/EC of the European Parliament and of the Council of 23 October 2000 establishing a framework for Community action in the field of water policy. Official Journal - European Communities Legislation 43.

EC, 2000b. The role of wetlands in the Water Framework Directive. Common implementation strategy for the Water Framework Directive 12.

Efficacitas, 2006. Estudio de Impacto Ambiental Definitivo: Proyecto Multiproposito BABA (in Spanish). Consorcio Hidroenergetico del Litoral, Guayaquil.

Endreny, T.A., Imbeah, N., 2009. Generating robust rainfall intensity-duration-frequency estimates with short-record satellite data. Journal of Hydrology 371, 182-191.

Falconi-Benitez, F., 2000. An integrated assessment of changes of land-use in Ecuador, Advances in Energy Studies, Porto Venere, Italy.

Feldman, A.D., 2000. Hydrologic Modeling System HEC-HMS, Technical Reference Manual, Feldman, A. ed. USACE, Davis, CA, USA.

Finlay, P.N., 1994. Introducing decision support systems. Blackwell Publishers, Oxford, UK.

Fleming, M.J., Doan, J.H., 2009. *Geospatial Hydrologic Modeling extension for HEC-HMS*. U.S. Army Corps of Engineers, Hydrologic Engineering Center, HEC, Washignton D.C., USA.

Geetha, K., Mishra, S.K., Eldho, T.I., Rastogi, A.K., Pandey, R.P., 2007. Modifications to SCS-CN Method for Long-Term Hydrologic Simulation. Journal of irrigation and drainage engineering. 133, 475.

Genova, K., Vassilev, V., Andonov, F., Vassileva, M., Konstantinova, S., 2004. A Multircirteria Analysis Decision Support System, International Conference on Computer System and Technologies, Rousse, Bulgaria, pp. 1-5.

Giupponi, C., Camera, R., Fassio, A., Lasut, A., Mysiak, J., Sgobbi, A., 2006. Network Analysis, Creative System Modelling and Decision Support: The NetSyMoD approach. Fondazione Eni Enrico Mattei 46.2006.

Giupponi, C., Cojocaru, G., Féas, J., Mysiak, J., Rosato, P., Zucca, A., 2010. mDSS, user's guide. Netsymod, Brussels, Belgium.

Goosen, H., Janssen, R., Vermaat, J., 2007. Decision support for participatory wetland decision-making. Ecological Engineering 30, 187-199.

Gorry, G.A., Scott Morton, M.S., 1971. A framework for management information systems. Massachusetts Institute of Technology, Cambridge, Mass, USA.

Greene, W., 2008. Econometric Analysis, 7th ed. Prentice Hall, New York, NY, USA.

Guetter, A.K., Georgakakos, K.P., Tsonis, A.A., 1996. Hydrologic applications of satellite data: 2. Flow simulation and soil water estimates. Journal of Geophysical Research 101, 26527-26538.

Gutjahr, A.L., Bras, R., 1993. Spatial variability in subsurface flow and transport: a review. Reliability Engineering & System Safety 42, 293-316.

Hattermann, F.F., Krysanova, V., Habeck, A., Bronstert, A., 2006. Integrating wetlands and riparian zones in river basin modelling. Ecological Modelling 199, 379-392.

Hein, T., Baranyi, C., Reckendorfer, W., Schiemer, F., 2004. The impact of surface water exchange on the nutrient and particle dynamics in side-arms along the River Danube, Austria. Science of the Total Environment. 328, 207.

Hein, T., Blaschke, A.P., Haidvogl, G., Hohensinner, S., Kucera-Hirzinger, S., Preiner, S., Reiter, K., Schuh, B., Weigelhofer, G., Zsuffa, I., 2005. Optimised management strategies for the Biosphere reserve Lobau - Austria - based on multicriteria decision support system, Vienna, Austria.

Hill, M.O., 1973. Diversity and Evenness: A Unifying Notation and Its Consequences. Ecology 54, 427-432.

Hodrick, R.J., Prescott, E.C., 1997. Postwar U.S. Business Cycles: An Empirical Investigation. Journal of Money, Credit and Banking 29, 1-16.

Huffman, G.J., Adler, R.F., Bolvin, D.T., Gu, G., Nelkin, E.J., Bowman, K.P., Hong, Y., Stocker, E.F., Wolff, D.B., 2007. The TRMM Multisatellite Precipitation Analysis (TMPA): Quasi-global, multiyear, combined-sensor precipitation estimates at fine scales. Journal of Hydrometeorology 8, 38-55.

Huffman, G.J., Adler, R.F., Bolvin, D.T., Nelkin, E.J., 2010. The TRMM Multi-Satellite Precipitation Analysis (TMPA), In: Gebremichael, M., Hossain, F. (Eds.), Satellite Rainfall Applications for Surface Hydrology. Springer Science, New York, USA.

Hunt, R.J., Welter, D.E., 2010. Taking Account of Unknown Unknowns. Ground Water Ground Water 48, 477-477.

Hwang, C.L., Yoon, K., 1981. Multiple attribute decision making : methods and applications. Springer-Verlag, Berlin, Germany; New York, USA.

INEC, 2002. Estadísticas del Censo de Población y Vivienda 2001 de Ecuador (in Spanish), from: www.inec.gob.ec. National Institute of Statistics & Census, Quito, Ecuador.

Janssen, R., Goosen, H., Verhoeven, M.L., Verhoeven, J.T.A., Omtzigt, A.Q., Maltby, E., 2005. Decision support for integrated wetland management. Environmental Modelling and Software 20, 215-229.

Jonoski, A., 2002. Hydroinformatics as Sociotechnology: Promoting Individual Stakeholder Participation by Using Network Distributed Decision Support Systems, Hydroinformatics and Knowledge Information Department. TU Delft / UNESCO-IHE, Delft.

Jonsdottir, J.F., Uvo, C.B., Clarke, R.T., 2008. Filling Gaps in Measured Discharge Series with Model-Generated Series. Journal of Hydrologic Engineering 13, 905-909.

Kaggwa, R.C., van Dam, A.A., Balirwa, J.S., Kansiime, F., Denny, P., 2008. Increasing fish production from wetlands at Lake Victoria, Uganda using organically manured seasonal wetland fish ponds. Wetlands ecology and management.

Keen, P.G.W., Scott Morton, M.S., 1978. Decision support systems : an organizational perspective. Addison-Wesley Pub. Co., Reading, Mass., USA.

Kent, D.M., 2001. Applied wetlands science and technology. Lewis Publishers, Boca Raton, FL, USA.

Kindsvater, C.E., Carter, R.W., 1959. Discharge characteristics of rectangular thin-plate weirs. American Society of Civil Engineers 24, 772-822.

King, R.G., Rebelo, S.T., 1999. Resuscitating Real Business Cycles. Handbook of Macroeconomics 1, 927-1007, Elsevier.

Kirk, J.A., Wise, W.R., Delfino, J.J., 2004. Water budget and cost-effectiveness analysis of wetland restoration alternatives: a case study of Levy Prairie, Alachua County, Florida. ECOLOGICAL ENGINEERING 22, 43-60.

Kone, B., Wymenga, E., al, e., 2002. Organisation socio-économique du delta intérieur du fleuve Niger (in French). Wetlands International, RIZA & Altenburg & Wymenga ecological consultants, Bamako (Mali), Lelystad & Veenwouden (NL).

Kottek, M., Grieser, J., Beck, C., Rudolf, B., Rubel, F., 2006. World Map of the Koppen-Geiger climate classification updated. Meteorological Journal 15, 259.

Kotze, D., Marneweck, G., Batchelor, A., Lindley, D., Collins, N., 2008. WET-EcoServices: A technique for rapidly assessing ecosystem services supplied by wetlands, In: Breen, C., Dini, J., Ellery, W., Mitchell, S., Uys, M. (Eds.), Wetland Management Series. Water Research Commission, Gezina, South Africa.

Krause, S., Jacobs, J., Bronstert, A., 2007. Modelling the impacts of land-use and drainage density on the water balance of a lowland-floodplain landscape in northeast Germany. Ecological Modelling 200, 475-492.

Krebs, C.J., 1978. Ecology : the experimental analysis of distribution and abundance. Harper & Row, New York.

Krysanova, V., Müller-Wohlfeil, D., Becker, A., 1998. Development and test of a spatially distributed hydrological/water quality model for mesoscale watersheds. Ecological Modelling 106, 261-289.

Küçük, M., Agiralioglu, N., 2006. Wavelet Regression Technique for Streamflow Prediction. Journal of Applied Statistics 33, 943-960.

Kummerow, C., Barnes, W., Kozu, T., Shiue, J., Simpson, J., 1998. TheTropical Rainfall Measuring Mission (TRMM) sensor package. Journal of Atmospheric and Oceanic Technology 15, 809-817.

La Jeunesse, I., Rounsevell, M., Vanclooster, M., 2003. Delivering a decision support system tool to a river contract: a way to implement the participatory approach principle at the catchment scale? Physics and Chemistry of the Earth, Parts A/B/C 28, 547-554.

Lascano, M., 2009. Programa Socio-Bosque: Ministry of Environment., Quito, Ecuador.

Lloyd, M., Ghelardi, R.J., 1964. A Table for Calculating the `Equitability' Component of Species Diversity. The Journal of Animal Ecology 33, 217-225.

Lovejoy, S.B., Lee, J.G., Randhir, T.O., Engel, B.A., 1997. Research needs for water quality management in the 21st century. Journal of Soil and Water conservation 52, 18-21.

Macfarlane, D., Kotze, D., Ellery, W., Walters, D., Koopman, V., Goodman, P., Goge, C., 2008. WET-Health: A technique for rapidly assessing wetland health, In: Breen, C., Dini, J., Ellery, W., Mitchell, S., Uys, M. (Eds.), Wetland Management Series. Water Research Commission, Gezina, South Africa.

Magurran, A.E., 1983. Ecological Diversity and its Measurement. Croom Helm, London, UK.

Malczewski, J., 1999. GIS and multicriteria decision analysis. J. Wiley & Sons, New York, USA.

Margalef, R., 1972. Homage to Evelyn Hutchinson, or why there is an upper limit to diversity. Transactions of the Conneticut Academy of Arts and Sciences 44, 211-235.

Matteau, M., Assani, A.A., Mesfioui, M., 2009. Application of multivariate statistical analysis methods to the dam hydrologic impact studies. Journal of Hydrology 371, 120-128.

Miller, R.C., Guertin, D.P., Heilman, P., 2004. Information Technology in Watershed Management Decision Making. Journal of the American Water Resources Association 40, 347-357.

Mitsch, W., Gosselink, J., 1986. Wetlands. Van Nostrand Reinhold, New York, USA.

Mitsch, W.J., Straskraba, M., Jorgensen, S.E., 1988. Wetland modelling, Amsterdam; New York.

Moliere, D., Lowry, J., Humphrey, C., 2009. Classifying the flow regime of data-limited streams in the wet-dry tropical region of Australia. Journal of Hydrology 367, 1-13.

Mynett, A.E., 2002. Environmental Hydroinformatics: The Way Ahead, Keynote address 5th International Conference on Hydroinformatics. IWA Publishing, Cardiff, UK, pp. 31-36.

Mynett, A.E., 2004. Artificial Intelligence Techniques in Environmental Hydroinformatics, Keynote Address at the International IAHR-Asian and Pacific Division Conference, Hong Kong, China.

Mynett, A.E., 2008. Environmental Hydroinformatics in Water Resources Research, Invited keynote for the International Symposium on Global Water Issues, 30th anniversary of the Water Resources Centre, Kyoto, Japan.

Nachtnebel, H.-P., 1994. Multicriterion decison making methods with ordinal and cardinal scales: ELECTRE I-III, In: Bogardi Janos J., Hans-Peter, N. (Eds.), Multicriteria decision analysis in water resources management. UNESCO, International Hydrological Programme, Paris, France.

Nash J.E., Sutcliffe J.V., 1970. River flow forecasting through conceptual models. Part I - A discussion of principles. Journal of Hydrology 10, 282-290.

National Assembly, 2008. Constitution of the Republic of Ecuador (in Spanish), published in the Official Register, October 20, 2010. Ministry of Foreign Affairs of Ecuador, Quito.

Nieto, J.J., 2007. Modelo de regresión lineal múltiple para determinar influencias del índice Niño 1+2 y la MJO sobre las precipitaciones en Guayaquil durante enero-febrero-marzo y abril (in Spanish). Acta Oceanográfica del Pacífico 14, 25-30.

Nieto, J.J., Martínez, R., Regalado, J., Hernández, F., 2002. Análisis de tendencia de series de tiempo oceanográficas y meteorológicas para determinar evidencias de Cambio Climático en la costa del Ecuador (in Spanish). Acta Oceanográfica del Pacífico 11, 17-21.

Noroña, B., 2009. Analysis of the decision making process in wetland management and the role of guidelines. Case study: Abras de Mantequilla - Prov. Los Rios, Ecuador, Water Management. UNESCO-IHE, Institute for Water Education, Delft, The Netherlands.

Olson, W.S., Kummerow, C.D., Heymsfield, G.M., Giglio, L., 1996. A Method for Combined Passive-Active Microwave Retrievals of Cloud and Precipitation Profiles. Journal of Applied Meteorology 35, 1763-1789.

Oram, B., Alcock, K., 2010. The Water Quality Index: Monitoring the Quality of Surfacewaters.

Oreskes, N., 2003. The role of quantitative models in science, In: Canham, C.D., Cole, J.J., Lauenroth, W.K. (Eds.), Models in Ecosystem Science. Princeton University Press, Princeton, NJ, USA.

Orlowsky, B., Gerstengarbe, F.W., Werner, P.C., 2008. A resampling scheme for regional climate simulations and its performance compared to a dynamical RCM. Theoretical and Applied Climatology 92, 209-223.

Paiva, R., Buarque, D.C., Collischonn, W., Sorribas, M., Allasia, D.G., Mendes, C., Tucci, C., Bonnet, M., 2011. Hydrologic and Hydrodynamic Modelling of the Amazon Basin using TRMM Rainfall Estimates. International Association of Hydrological Sciences (IAHS) 343, 72-77.

Pan, F., Pachepsky, Y.A., Guber, A.K., McPherson, B.J., Hill, R.L., 2012. Scale effects on information theory-based measures applied to streamflow patterns in two rural watersheds. Journal of Hydrology 414-415, 99-107.

Phillips, P., 2010. The Mysteries of Trend, Cowles Foundation for Research in Economics. Yale University, New Haven, Connecticut, USA.

Prado, M., Macías, P., Cajas, J., Elias, E., Revelo, W., 2004. Prediagnóstico de las condiciones físicas, químicas y biológicas en el sistema fluvial de la provincia de Los Rios (in Spanish). National Fishery Institute (INP), Guayaquil, Ecuador.

Price, R.K., Vojinovic, Z., 2011. Urban hydroinformatics : data, models, and decision support for integrated urban water management. IWA Publishing, London, UK.

Ramsar Secretariat, 2010. Manejo de cuencas hidrográficas: Integración de la conservación y del uso racional de los humedales en el manejo de las cuencas hidrográficas (in Spanish), In: 4th_edition (Ed.), Ramsar Manuals for the rational use of wetlands. Ramsar Convention, Gland, Switzerland.

Ravn, M., Uhlig, H., 2002. On adjusting the Hodrick-Prescott filter for the frequency of observations. The Review of Economics and Statistics 84, 371-375.

Razzak, W., 1997. The Hodrick-Prescott technique: A smoother versus a filter: An application to New Zealand GDP. Economics Letters 57, 163-168.

Reitsma, R.F., 1990. Functional Classification of Space. Aspects of Site suitability Assessment in a Decision Support Environment. International Institute for applied systems analysis, Laxenburg, Austria.

Rizzoli, A.E., Young, W.J., 1997. Delivering environmental decision support systems: software tools and techniques. Environmental Modelling & Software 12, 237-249.

Rollenbeck, R., Bendix, J., 2011. Rainfall distribution in the Andes of southern Ecuador derived from blending weather radar data and meteorological field observations. Atmospheric Research 99, 277-289.

Romero, P., Jiménez, S., Herrera, J., Cabrera, J., Garcés, D., 2009. Estudio Hidrogeológico del Humedal Abras de Mantequilla (in Spanish). ESPOL, Guayaquil, Ecuador, p. 22pp.

Sanborn, S.C., Bledsoe, B.P., 2006. Predicting streamflow regime metrics for ungauged streamsin Colorado, Washington, and Oregon. Journal of Hydrology 325, 241-261.

Schlicht, E., 2004. Estimating the smoothing parameter in the Hodrick-Prescott Filter. Munich Discussion Paper, Universität München, Department of Economics 2.

SEI, 2009. WEAP (Water Evaluation And Planning System) tutorial;, Stockholm Environment Institute, Stockholm, Sweden.

Sen, P., Yang, J.-B., 1998. Multiple criteria decision support in engineering design. Springer, London, UK; New York, USA.

Sharffenberg, W.A., Fleming, M.J., 2010. Hydrologic Modeling System, HEC-HMS User's Manual. U.S. Army Corps of Engineers (Hydrologic Engineering Center -HEC), Washignton D.C., USA.

Simon, H.A., 1981. The sciences of the artificial. MIT Press, Cambridge, Mass., USA.

Simonovic, S.P., 2009. Managing Water Resources: Methods and Tools for a Systems Approach. UNESCO Publishing, Paris, France / London, UK.

Simpson, J., Adler, R.F., North, G.R., 1988. A proposed Tropical Rainfall Measuring Mission (TRMM) satellite. Bulletin of American Meteorological Society 69, 278-295.

Skaags, R.W., Khaleel, R., 1982. Infiltration, Hydrologic modeling of small watersheds. American Society of Agricultural Engineers, St. Joseph, MI, USA.

Smith, L., Turcotte, D., Isacks, B., 1998. Stream flow characterization and feature detection using a discrete wavelet transform. Hydrological Processes 12, 233-249.

Southgate, D., Whitaker, M., 1994. Economic Progress and the Environment: One Developing Country's Policy Crisis. Oxford University Press.

Sprague, R.H., Watson, H.J., 1993. Decision support systems : putting theory into practice. Prentice Hall, Englewood Clifts, N.J.

Tecle, A., Duckstein, L., 1994. Concepts of Multicriterion Decision Making, In: Bogardi Janos J., Hans-Peter, N. (Eds.), Multicriteria decision analysis in water resources management. UNESCO, International Hydrological Programme, Paris, France.

Tian, Y., Peters-Lidard, C.D., 2010. A global map of uncertainties in satellite-based precipitation measurements. Geophysical Research Letters 37, doi:10.1029/2010GL046008.

Turban, E., 2007. Decision support and business intelligence systems. Pearson Prentice Hall, Upper Saddle River, N.J., USA.

USDA, 1986. *TR-55: Urban Hydrology for small watersheds*. US Department of Agriculture; Nacional Resources Conservation Service (NCRS), Washignton.

USDA, 2004. National Engineering Handbook, part 630: Hydrology: Estimation of Direct Runoff from Storm Rainfall. Natural Resources Conservation Service (NRCS), US Department of Agriculture, Washington, DC, USA.

van Ast, A.J., 2000. Interactive management of international river basins; experiences in Northern America and Western Europe. Physics and Chemistry of the Earth, Part B: Hydrology, Oceans and Atmosphere 25, 325-328.

van Dam, A.A., Kelderman, P., Kansiime, F., Dardona, A., 2007. A simulation model for nitrogen retention in a papyrus wetland near Lake Victoria, Uganda (East Africa). Wetlands ecology and management 15, 469-480.

van der Knijff, J.M., Younis, J., de Roo, A.P.J., 2010. LISFLOOD: a GIS-based distributed model for river basin scale water balance and flood simulation. International Journal of Geographical Information Science 24, 189-212.

van Griensven, A., Alvarez-Mieles, M., 2009. Environmental monitoring in The Abras de Mantequilla Wetland & Area of Influence. UNESCO-IHE, Delft, The Netherlands.

van Griensven, A., Xuan, Y., Haguma, D., Niyonzima, W., 2008. Understanding riverine wetland-catchment processes using remote sensing data and modeling, In: Sánchez-Marrè, M., Béjar, J., Comas, J., Rizzoli, A.E., Guariso, G. (Eds.), International Congress on Environmental Modelling and Software. iEMSs, Barcelona, Spain, pp. 462-469.

Vernimmen, R.R.E., Hooijer, A., Mamenun, Aldrian, E., van Dijk, A.I.J.M., 2012. Evaluation and bias correction of satellite rainfall data for drought monitoring in Indonesia. Journal of Hydrology and Earth System Sciences 16, 133-146.

Villa-Cox, G., Arias-Hidalgo, M., Mino, S., Delgado-Cabrera, L., 2011. Scenario descriptions, Management options and asociated indicators: Fact-Sheet for the Abras de Mantequilla case study, WP7. WETWin project, ESPOL University, Guayaquil, Ecuador.

Vuille, M., Bradley, R., Keimig, F., 2000. Climate Variability in the Andes of Ecuador and Its Relation to Tropical Pacific and Atlantic Sea Surface Temperature Anomalies. Journal of Climate 13, 2520-2535.

Waite, P.J., 1982. Competition for water resources of the Rio Guayas, Ecuador, Optimal Allocation of Water Resources, Proceedings of the Exeter Symposium. IAHS, Exeter, UK, p. 135.

Walker, J.S., 1999. A primer on wavelets and their scientific applications. Chapman & Hall/CRC, Boca Raton, Fla.

Walters, D., Shrubsole, D., 2003. Agricultural drainage and wetland management in Ontario. Journal of Environmental Management 69, 369-379.

Wang, W., Hu S., Y., L., 2011. Wavelet Transform Method for Synthetic Generation of Daily Streamflow. Water Resources Management, 41-57.

Wang, X., Li, L., Lockington, D., Pullar, D., Jeng, D.-S., 2005. Self-Organizing Polynomial Neural Network for Modelling Complex Hyrological Processes. The University of Sydney, Sydney, Australia, p. 30pp.

Wattenbach, M., 2008. The hydrological effects of cahnges in forest area and species composition in the federal state of Brandenburg, Germany, Department of Mathematics and Natural Sciences. University of Potsdam, Potsdam, Germany.

Welp, M., 2001. The use of decision support tools in participatory river basin management. Physics and Chemistry of the Earth, Part B: Hydrology, Oceans and Atmosphere 26, 535-539.

Wilheit, T.T., 1988. Error analysis for the Tropical Rainfall Measuring Mission, In: Theon, J.S., Fugono, N. (Eds.), Tropical Rainfall Measurements, Hampton, Va, USA.

Williams, M., 2001. Conservation of Wetlans, International Encyclopedia of the Social & Behavioral Sciences. Elsevier Science Ltd., pp. 2621-2624.

Witlox, F., 2005. Expert systems in land-use planning: An overview. Expert Systems with Applications 29, 437-445.

Wong, W.F., Chiu, L.S., 2008. Spatial and Temporal Analysis of Rain Gauge Data and TRMM Rainfall Retrievals in Hong Kong. Geographic Information Sciences 14, 105-112.

Yilmaz, K.K., Gupta, H., Hogue, T.S., Hsu, K., Wagener, T., Sorooshian, S., 2005. Evaluating the utlility of satellite-based precipitation estimates for runoff prediction in ungauged basins. Regional Hydrological Impacts of Climatic Change - Impact Assessment and Decision Making, Proceedings of the VII IAHS Meeting. IAHS, Foz do Iguaçu, Brazil.

Yoon, K., Hwang, C.L., 1995. Multiple attribute decision making : an introduction. Sage Publications, Thousand Oaks, CA, USA.

Zacharias, I., Dimitriou, E., Koussouris, T., 2005. Integrated water management scenarios for wetland protection: application in Trichonis Lake. Environmental Modelling & Software 20, 177-185.

Zsuffa, I., 2008. WETWIN Project, Part B: Enhancing the role of wetlands in integrated water resources management for twinned river basins in EU, Africa and South-America in support of EU Water Initiatives, In: VITUKI (Ed.). European Union, Budapest, Hungary.

Zsuffa, I., Cools, J., 2011. Drivers - States - Impacts - Responses: DSIR analyses at the study sites, D3.2, v31. The WETwin project, 7th Framework Programme.

Zwarts, L., Diallo, M., 2002. Eco-hydrologie du Delta, In: Wymenga, E., Kone, B., Zwarts, L. (Eds.), Le delta intérieur du fleuve Niger: ecologie et gestion durable des resources naturelles (in French). Wetlands International, RIZA & Altenburg & Wymenga ecological consultants, Bamako (Mali), Lelystad & Veenwouden (NL).

# Appendix A:

# WETWin Project

# Questionnaire for the evaluation of expert elicitation

# Abras de Mantequilla wetland case study

**Technical staff**

*Land Use and Agricultural practices*
Gonzalo Villa-Cox

*Hydrology, modelling*
Mijail Arias-Hidalgo

*Ecology and biodiversity*
Sandra Mino-Quezada
Gabriela Alvarez-Mieles

*Institutional issues and sociology*
Roberto Sáenz

## A.1 Introduction for the questionnaire

A description on the project baseline has been presented to the audience, consisting in the following components:

- Characterization of the hydrological regime of the Chojampe River subcatchment.
- Current landuse and soil type in the Chojampe subcatchment.
- Current agricultural practices in the study area.
- Characterization of the current status of ecology and biodiversity.

Afterwards, the climate change projections were described as well as the major infrastructure works that currently SENAGUA is implementing. These two elements were the basis for the Business As Usual scenario (BAU). Since these factors are exogenous to the project (beyond its control), they are regarded as the situation that most likely the wetland and its surroundings will cope with in case no management action is taken during a time span of around 30 years or less.

The rainfall time series that served to analyze the climate changes (BAU scenario) were built based on 0.5 degree variations (air temperature) projected by the PIK Institute (in Germany). According to Nieto et al. (2002), these variations tend to be positive (0.017 °C/year), may resulting thus in precipitation increments. The projected rainfall was inserted in a HEC-HMS model (built for the baseline) to obtain headflows for each river involved in the water allocation model WEAP. As the time span lasted around 30-40 years, it was divided in three sub-periods to facilitate the simulations in WEAP:

- Representative year A: period 2007-2010 (1$^{st}$ "decade")
- Representative year B: period 2011-2020
- Representative year C: period 2021-2030
- Representative year D: period 2031-2040

Once the BAU was described, five management options were proposed, as follows:

- O1 – Increase the storage capacity of the Abras de Mantequilla wetland via the use of hydraulic gates at the connection point with the Nuevo River.
- O2 – The adoption of an integrated plan to ameliorate the agricultural practices at a local scale: This includes the prohibition of red and yellow pesticides, the reduction of fertilizers, with a rate of 10% of each crop area per decade.
- O3 – Crop substitution, replacing short-term crops (corn and beans mainly) for long-term crops (cocoa, passion fruit and other fruit trees). This will be carried out via the potential adoption of agroforestry practices, with a rate of 10% of the total crop surface per decade.
- O4 – Crop substitution, replacing short-term crops (corn and beans mainly) for long-term crops (cocoa, passion fruit and other fruit trees). This will be carried

out via the potential adoption of more intensive agroforestry practices, with a rate of 20% of the total crop surface per decade.

* O5 – Implementation of ecological corridors to recover the forest patches and thus preserve and enhance the local biodiversity. The rate is 5% of the total surface each decade.

The goal of the following questionnaire is to measure, based on your expert criteria and the technical material previously shown, the impact that each of the management options (within the WETWin project framework) have on the different socio-economic, environmental and institutional issues. In this regard, the "management solutions" (MS) is referred to different combinations of the management options described above, in similar way as they would have been arranged in an integrated management plan. To this end, it is requested that you evaluate a set of criteria, using the Lickert scale, under each of the management solution that will be presented. These MS are as follows:

* MS0 – BAU
* MS1 – O1 + O2
* MS2 – O1 + O2 + O3
* MS3 – O1 + O2 + O4
* MS4 – O1 + O2 + O3 + O5
* MS5 – O1 + O2 + O4 + O5

This means that for every management solution you are asked to consider the joint impact of the incremented water storage and the improvement of agricultural practices at a local scale. Both by itself and for the combination of it with different schemes of landuse changes (crop substitution and ecological corridors).

For each the socio-economic/environmental indicators, you are asked to assess the relative impact that each solution may produce on the criterion referred by each indicator, in comparison with the MS0 (BAU). All indicators are scaled between 1 and 5 where 3 mean equivalence with respect to the BAU or that the solution has no effect on the criterion. Values higher than 3 means improvement and values lesser than 3 means a negative impact, with respect to the BAU.

The institutional indicators aim to collect an absolute criterion on the capacity of the management structure (local and regional/national) and the local stakeholders (local dwellers) to adopt the diverse management solutions. Local management structures are the local authorities (e.g. municipalities, parishes juntas, etc) directly responsible and involved in the wetland management; whereas regional/national management structure is referred to the central government and private entities (Ministries, provincial councils, NGOs, etc) whose role within the management is to provide support and to determine high-order policies for water & environmental resources management in Ecuador. Like for the other indicators, the institutional ones are also measured from 1 to 5, where the highest values mark a better eagerness or capacity. However, this time the evaluation is not carried out with respect to the BAU; rather what is sought is an absolute measure of the capacity of implementation. This is why it is relevant the values that these indicators take under the MS0 (BAU). Therefore, you will be asked to evaluate the institutional indicators under the MS0 (compulsorily), which means the capacity of the management structure and local stakeholders to cope with the implications of the BAU scenario without adopting any of the management solutions proposed by the WETWin project. In this regard, the category #3 marks an adequate capacity to adopt a management solution. Finally, it is worth emphasizing that for each indicator in this survey, a special category (0) has been added. Should the criterion that the indicator pursues is not within your specialization please make use of this option.

## A.2 Instructions for the questionnaire

The survey is divided in three sections (i) socioeconomic / environmental; (ii) ecological/biodiversity; and (iii) institutional indicators. You will find, for each indicator, one table whose rows represent indicator categories and whose columns are the management solutions, as were described above. On the left you will find a description of each category in the context of the indicator.

For the first two sections, you are asked to evaluate the category that corresponds to each column (MS) according to what you consider appropriate as measurement. Please, mark with an X to do so. Noteworthy is that in any case it is valid to mark one column (MS) more than once (two or more categories simultaneously). Herewith, an example of a correctly filled matrix for a given indicator:

| Value/MS | MS1 | MS2 | MS3 | MS4 | MS5 |
|----------|-----|-----|-----|-----|-----|
| 5 | X | | | | |
| 4 | | | X | | |
| 3 | | | | | X |
| 2 | | X | | | |
| 1 | | | | X | |
| 0 | | | | | |

Please note that you must not leave any column in blank. In case you choose 0, you must do it in the following way to validate your answer:

| Value/MS | MS1 | MS2 | MS3 | MS4 | MS5 |
|----------|-----|-----|-----|-----|-----|
| 5 | | | | | |
| 4 | | | | | |
| 3 | | | | | |
| 2 | | | | | |
| 1 | | | | | |
| 0 | X | X | X | X | X |

**Expert's name:**     Paul Herrera, PhD.
**Institution:**         ESPOL University – Guayaquil, Ecuador
**Position:**            Director, Rural Research Center

# I. Socio-economic & environmental indicators

## 1.   Crop investment / sowing costs per Ha.

| Category | Description | Value/MS | MS1 | MS2 | MS3 | MS4 | MS5 |
|---|---|---|---|---|---|---|---|
| Significant improvement | Significant costs reduction | 5 | | | | | |
| Small improvement | Small costs reduction | 4 | X | | | | |
| Indifferent to BAU | Same as current cost structure | 3 | | X | | | |
| Small detriment | Small increment in costs | 2 | | | X | | |
| Significant detriment | Significant increment in costs | 1 | | | | X | X |
| Not defined criterion | It is not my field of specialization | 0 | | | | | |

## 2.   Crop maintenance costs per Ha.

| Category | Description | Value/MS | MS1 | MS2 | MS3 | MS4 | MS5 |
|---|---|---|---|---|---|---|---|
| Significant improvement | Significant costs reduction | 5 | | | | | |
| Small improvement | Small costs reduction | 4 | X | | | | |
| Indifferent to BAU | Same as current cost structure | 3 | | X | | | |
| Small detriment | Small increment in costs | 2 | | | X | | |
| Significant detriment | Significant increment in costs | 1 | | | | X | X |
| Not defined criterion | It is not my field of specialization | 0 | | | | | |

## 3.   Food security, local scale

| Category | Description | Value/MS | MS1 | MS2 | MS3 | MS4 | MS5 |
|---|---|---|---|---|---|---|---|
| Significant improvement | Broader crop diversification / significant improvement in local food security | 5 | | | | X | X |
| Small improvement | Broader crop diversification / small improvement in local food security | 4 | | X | X | | |
| Indifferent to BAU | Current status of local food security (diversification of short-term & long-term crops) | 3 | X | | | | |
| Small detriment | Narrower crop diversification / small impact in local food security | 2 | | | | | |
| Significant detriment | Crop specialization / significant impact in local food security | 1 | | | | | |
| Not defined criterion | It is not my field of specialization | 0 | | | | | |

## 4. Crop productivity per Ha

| Category | Description | Value/MS | MS1 | MS2 | MS3 | MS4 | MS5 |
|---|---|---|---|---|---|---|---|
| Significant improvement | Significant increment in productivity | 5 | | | | X | X |
| Small improvement | Small increment in productivity | 4 | | X | X | | |
| Indifferent to BAU | Same productivity as current conditions | 3 | X | | | | |
| Small detriment | Small decrement in productivity | 2 | | | | | |
| Significant detriment | Significant decrement in productivity | 1 | | | | | |
| Not defined criterion | It is not my field of specialization | 0 | | | | | |

## 5. Input of sediment and nutrients due to landuse

| Category | Description | Value/MS | MS1 | MS2 | MS3 | MS4 | MS5 |
|---|---|---|---|---|---|---|---|
| Significant improvement | Significant decrement of sediment and nutrient inputs | 5 | | | | | X |
| Small improvement | Small decrement of sediment and nutrient inputs | 4 | | | | X | |
| Indifferent to BAU | Current level of sediment and nutrient inputs | 3 | X | X | X | | |
| Small detriment | Small increment of sediment and nutrient inputs | 2 | | | | | |
| Significant detriment | Significant increment of sediment and nutrient inputs | 1 | | | | | |
| Not defined criterion | It is not my field of specialization | 0 | | | | | |

## 6. Income level of local population

| Category | Description | Value/MS | MS1 | MS2 | MS3 | MS4 | MS5 |
|---|---|---|---|---|---|---|---|
| Significant improvement | Significant improvement of income level | 5 | | | | X | X |
| Small improvement | Small improvement of income level | 4 | | X | X | | |
| Indifferent to BAU | Current level of income level | 3 | X | | | | |
| Small detriment | Small decrement of income level | 2 | | | | | |
| Significant detriment | Significant decrement of income level | 1 | | | | | |
| Not defined criterion | It is not my field of specialization | 0 | | | | | |

## 7. Economical associability potential of local stakeholders

| Category | Description | Value/MS | MS1 | MS2 | MS3 | MS4 | MS5 |
|----------|-------------|----------|-----|-----|-----|-----|-----|
| Significant improvement | The alternative remarkably enhances the potential of association | 5 | | | | X | X |
| Small improvement | The alternative slightly enhances the potential of association | 4 | | X | X | | |
| Indifferent to BAU | The alternative does not affect the potential of association | 3 | X | | | | |
| Small detriment | The alternative slightly damages the potential of association | 2 | | | | | |
| Significant detriment | The alternative remarkably damages the potential of association | 1 | | | | | |
| Not defined criterion | It is not my field of specialization | 0 | | | | | |

## 8. Navigability inside the wetland

| Category | Description | Value/MS | MS1 | MS2 | MS3 | MS4 | MS5 |
|----------|-------------|----------|-----|-----|-----|-----|-----|
| Significant improvement | Navigability increases remarkably | 5 | | | | | |
| Small improvement | Navigability slightly increases | 4 | | | | | |
| Indifferent to BAU | Current navigability is not affected | 3 | | | | | |
| Small detriment | Navigability slightly decreases | 2 | | | | | |
| Significant detriment | Navigability decreases remarkably | 1 | | | | | |
| Not defined criterion | It is not my field of specialization | 0 | X | X | X | X | X |

## 9. Touristic potential at AdM

| Category | Description | Value/MS | MS1 | MS2 | MS3 | MS4 | MS5 |
|----------|-------------|----------|-----|-----|-----|-----|-----|
| Significant improvement | Touristic potential increases remarkably | 5 | | | | | X |
| Small improvement | Touristic potential slightly increases | 4 | | | | X | |
| Indifferent to BAU | Touristic potential is not affected | 3 | X | X | X | | |
| Small detriment | Touristic potential slightly decreases | 2 | | | | | |
| Significant detriment | Touristic potential increases remarkably | 1 | | | | | |
| Not defined criterion | It is not my field of specialization | 0 | | | | | |

## 10.   Educational level / Environmental awareness

| Category | Description | Value/MS | MS1 | MS2 | MS3 | MS4 | MS5 |
|---|---|---|---|---|---|---|---|
| Significant improvement | The alternative promotes a significant improvement of the educational level & environmental awareness | 5 | | | | X | X |
| Small improvement | The alternative promotes a small improvement of the educational level & environmental awareness | 4 | | X | X | | |
| Indifferent to BAU | The alternative does not affect the current educational level & environmental awareness | 3 | X | | | | |
| Small detriment | The alternative causes a small reduction of the educational level & environmental awareness | 2 | | | | | |
| Significant detriment | The alternative causes a significant reduction of the educational level & environmental awareness | 1 | | | | | |
| Not defined criterion | It is not my field of specialization | 0 | | | | | |

## II. Ecological and biodiversity indicators

### 1.   Eutrophication

| Category | Description | Value/MS | MS1 | MS2 | MS3 | MS4 | MS5 |
|---|---|---|---|---|---|---|---|
| Significant improvement | Significant reduction on water hyacinth coverage | 5 | | | | | |
| Small improvement | Small reduction on water hyacinth coverage | 4 | | | | | |
| Indifferent to BAU | Current average water hyacinth coverage | 3 | | | | | |
| Small detriment | Small increment on water hyacinth coverage | 2 | | | | | |
| Significant detriment | Significant increment on water hyacinth coverage | 1 | | | | | |
| Not defined criterion | It is not my field of specialization | 0 | X | X | X | X | X |

### 2.   Biodiversity

| Category | Description | Value/MS | MS1 | MS2 | MS3 | MS4 | MS5 |
|---|---|---|---|---|---|---|---|
| Significant improvement | Significant increment on biodiversity | 5 | | | | X | X |
| Small improvement | Small increment on biodiversity | 4 | | | | | |
| Indifferent to BAU | Current level of biodiversity | 3 | X | X | X | | |
| Small detriment | Slight loss of biodiversity | 2 | | | | | |
| Significant detriment | Significant loss of biodiversity | 1 | | | | | |
| Not defined criterion | It is not my field of specialization | 0 | | | | | |

## III. Institutional indicators

### 1. Capacity of local stakeholders to adopt a management solution

| Category | Description | Value | MS0 | MS1 | MS2 | MS3 | MS4 | MS5 |
|---|---|---|---|---|---|---|---|---|
| Optimal | The capacity of local stakeholders is optimal to adopt the alternative. | 5 | | | | | | |
| Adequate | The capacity of local stakeholders can well face the alternative challenges, however it is still prone for enhancement | 4 | | | | | X | X |
| Acceptable | The capacity of local stakeholders can still face the alternative challenges, however it is necessary to reinforce that capacity in an adequate manner | 3 | | | X | X | | |
| Deficient | The capacity of local stakeholders is deficient and requires major improvements to adopt the alternative. | 2 | X | X | | | | |
| Not suitable | There might not be sufficient capacity to adopt the alternative and it is not possible to improve it | 1 | | | | | | |
| Not defined criterion | It is not my field of specialization | 0 | | | | | | |

### 2. Local management structure capacity to adopt a management solution

| Category | Description | Value | MS0 | MS1 | MS2 | MS3 | MS4 | MS5 |
|---|---|---|---|---|---|---|---|---|
| Optimal | Management and coordination capacity of the regional management structure is excellent to adopt the alternative | 5 | | | | | | |
| Adequate | Management and coordination capacity of the regional management structure can appropriately face the measure, however still prone for enhancement | 4 | | | | | X | X |
| Acceptable | Management and coordination capacity of the regional management structure can still face the measure, but requires strengthen to do it adequately | 3 | | | X | X | | |
| Deficient | Management and coordination capacity of the regional management structure is deficient; requires major changes to adopt the alternative | 2 | X | X | | | | |
| Not suitable | There might not be sufficient management and coordination capacity to implement the alternative and it is not possible to do improvements. | 1 | | | | | | |
| Not defined criterion | It is not my field of specialization | 0 | | | | | | |

## 3. Regional/national Management structure capacity to negotiate support and coordinate actions

| Category | Description | Value | MS0 | MS1 | MS2 | MS3 | MS4 | MS5 |
|---|---|---|---|---|---|---|---|---|
| Optimal | Management and coordination capacity of the regional management structure is excellent to adopt the alternative | 5 | | | | | | |
| Adequate | Management and coordination capacity of the regional management structure can appropriately face the measure, however still prone for enhancement | 4 | | | | | X | X |
| Acceptable | Management and coordination capacity of the regional management structure can still face the measure, but requires strengthen to do it adequately | 3 | | X | X | X | | |
| Deficient | Management and coordination capacity of the regional management structure is deficient; requires major changes to adopt the alternative | 2 | | | | | | |
| Not suitable | There might not be sufficient management and coordination capacity to implement the alternative and it is not possible to do improvements. | 1 | X | | | | | |
| Not defined criterion | It is not my field of specialization | 0 | | | | | | |

## 4. Budget adequacy to undertake Management Solutions

| Category | Description | Value | MS0 | MS1 | MS2 | MS3 | MS4 | MS5 |
|---|---|---|---|---|---|---|---|---|
| Significant increment | The current local budget structure is adequate but it requires a significant increment on the budget amount to adopt an alternative | 5 | | | | | X | X |
| Low increment | The current local budget structure is adequate but it requires a small increment on the budget amount to adopt an alternative | 4 | | | | X | | |
| Optimal structure | The current local budget structure is adequate and an budget increment is not needed to adopt an alternative | 3 | | X | X | | | |
| Reform + low increment | The current local budget structure must be reformed and it requires a small increment on the budget amount to adopt an alternative | 2 | | | | | | |
| Reform + significant increment | The current local budget structure must be reformed and it requires a significant increment on the budget amount to adopt an alternative | 1 | X | | | | | |
| Not defined criterion | It is not my field of specialization | 0 | | | | | | |

# Appendix B:

# Socioeconomic & Institutional indicators

| | Indicator (& code) | Category | Description | Lickert Scale $a_l$ |
|---|---|---|---|---|
| **S O C I O E C O N O M I C   I N D I C A T O R S** | Crop investment / sowing costs per Ha. (crop_scost) | Significant improvement | Significant costs reduction | 5 |
| | | Small improvement | Small costs reduction | 4 |
| | | Indifferent to BAU | Same as current cost structure | 3 |
| | | Small detriment | Small increment in costs | 2 |
| | | Significant detriment | Significant increment in costs | 1 |
| | Crop maintenance costs per Ha. (crop_acost) | Significant improvement | Significant costs reduction | 5 |
| | | Small improvement | Small costs reduction | 4 |
| | | Indifferent to BAU | Same as current cost structure | 3 |
| | | Small detriment | Small increment in costs | 2 |
| | | Significant detriment | Significant increment in costs | 1 |
| | Food security, local scale (food_safe) | Significant improvement | Broader crop diversification / significant improvement in local food security | 5 |
| | | Small improvement | Broader crop diversification / small improvement in local food security | 4 |
| | | Indifferent to BAU | Current status of local food security (diversification of short-term & long-term crops) | 3 |

| | | | | |
|---|---|---|---|---|
| **S O C I O E C O N O M I C   I N D I C A T O R S** | | Small detriment | Narrower crop diversification / small impact in local food security | 2 |
| | | Significant detriment | Crop specialization / significant impact in local food security | 1 |
| | Crop productivity per Ha. (crop_prod) | Significant improvement | Significant increment in productivity | 5 |
| | | Small improvement | Small increment in productivity | 4 |
| | | Indifferent to BAU | Same productivity as current conditions | 3 |
| | | Small detriment | Small decrement in productivity | 2 |
| | | Significant detriment | Significant decrement in productivity | 1 |
| | Input of sediment and nutrients due to landuse (sed_nutr) | Significant improvement | Significant decrement of sediment and nutrient inputs | 5 |
| | | Small improvement | Small decrement of sediment and nutrient inputs | 4 |
| | | Indifferent to BAU | Current level of sediment and nutrient inputs | 3 |
| | | Small detriment | Small increment of sediment and nutrient inputs | 2 |
| | | Significant detriment | Significant increment of sediment and nutrient inputs | 1 |
| | Income level of local population (income_lev) | Significant improvement | Significant improvement of income level | 5 |
| | | Small improvement | Small improvement of income level | 4 |
| | | Indifferent to BAU | Current level of income level | 3 |
| | | Small detriment | Small decrement of income level | 2 |
| | | Significant detriment | Significant decrement of income level | 1 |
| | Economical associability potential of local stakeholders (econ_asoc) | Significant improvement | The alternative remarkably enhances the potential of association | 5 |
| | | Small improvement | The alternative slightly enhances the potential of association | 4 |

| | | | | |
|---|---|---|---|---|
| **S O C I O E C O N O M I C   I N D I C A T O R S** | | Indifferent to BAU | The alternative does not affect the potential of association | 3 |
| | | Small detriment | The alternative slightly damages the potential of association | 2 |
| | | Significant detriment | The alternative remarkably damages the potential of association | 1 |
| | Navigability (wtr_navig) | Significant improvement | Navigability increases remarkably | 5 |
| | | Small improvement | Navigability slightly increases | 4 |
| | | Indifferent to BAU | Current navigability is not affected | 3 |
| | | Small detriment | Navigability slightly decreases | 2 |
| | | Significant detriment | Navigability decreases remarkably | 1 |
| | Touristic potential at AdM (touris_pot) | Significant improvement | Touristic potential increases remarkably | 5 |
| | | Small improvement | Touristic potential slightly increases | 4 |
| | | Indifferent to BAU | Touristic potential is not affected | 3 |
| | | Small detriment | Touristic potential slightly decreases | 2 |
| | | Significant detriment | Touristic potential increases remarkably | 1 |
| | Educational level / Environmental awareness (envir_educ) | Significant improvement | The alternative promotes a significant improvement of the educational level & environmental awareness | 5 |
| | | Small improvement | The alternative promotes a small improvement of the educational level & environmental awareness | 4 |
| | | Indifferent to BAU | The alternative does not affect the current educational level & environmental awareness | 3 |
| | | Small detriment | The alternative causes a small reduction of the educational level & environmental awareness | 2 |

| Indicator | Category | Description | Lickert Scale $a_j$ |
|---|---|---|---|
| | Significant detriment | The alternative causes a significant reduction of the educational level & environmental awareness | 1 |
| Capacity of local stakeholders to adopt a management solution (stak_capac) | Optimal | The capacity of local stakeholders is optimal to adopt the alternative. | 5 |
| | Adequate | The capacity of local stakeholders can well face the alternative challenges, however it is still prone for enhancement | 4 |
| | Acceptable | The capacity of local stakeholders can still face the alternative challenges, however it is necessary to reinforce that capacity in an adequate manner | 3 |
| | Deficient | The capacity of local stakeholders is deficient and requires major improvements to adopt the alternative. | 2 |
| | Not suitable | There might not be sufficient capacity to adopt the alternative and it is not possible to improve it | 1 |
| Local management structure capacity to adopt an alternative (lcmgt_cap) | Optimal | Management and coordination capacity of the regional management structure is excellent to adopt the alternative | 5 |
| | Adequate | Management and coordination capacity of the regional management structure can appropriately face the measure, however still prone for enhancement | 4 |

Note: Left vertical label: INSTITUTIONAL INDICATORS

| | | | | |
|---|---|---|---|---|
| I N S T I T U T I O N A L   I N D I C A T O R S | | Acceptable | Management and coordination capacity of the regional management structure can still face the measure, but requires strengthen to do it adequately | 3 |
| | | Deficient | Management and coordination capacity of the regional management structure is deficient; requires major changes to adopt the alternative | 2 |
| | | Not suitable | There might not be sufficient management and coordination capacity to implement the alternative and it is not possible to do improvements. | 1 |
| | Regional/national management structure capacity to negotiate support and coordinate actions (rgmgt_cap) | Optimal | Management and coordination capacity of the regional management structure is excellent to adopt the alternative | 5 |
| | | Adequate | Management and coordination capacity of the regional management structure can appropriately face the measure, however still prone for enhancement | 4 |
| | | Acceptable | Management and coordination capacity of the regional management structure can still face the measure, but requires strengthen to do it adequately | 3 |
| | | Deficient | Management and coordination capacity of the regional management structure is deficient; requires major changes to adopt the alternative | 2 |

| | | Not suitable | There might not be sufficient management and coordination capacity to implement the alternative and it is not possible to do improvements. | 1 |
|---|---|---|---|---|
| | Budget adequacy to undertake Management Solutions. | Significant increment | The current local budget structure is adequate but it requires a significant increment on the budget amount to adopt an alternative | 5 |
| | | Low increment | The current local budget structure is adequate but it requires a small increment on the budget amount to adopt an alternative | 4 |
| | | Optimal structure | The current local budget structure is adequate and an budget increment is not needed to adopt an alternative | 3 |
| | | Reform + low increment | The current local budget structure must be reformed and it requires a small increment on the budget amount to adopt an alternative | 2 |
| | | Reform + significant increment | The current local budget structure must be reformed and it requires a significant increment on the budget amount to adopt an alternative | 1 |

# About the author

Mijail Arias-Hidalgo was born in Guayaquil, Ecuador on January 20, 1979. He followed his elementary and middle school studies in the same city. In June 2002 he obtained a Bachelor degree in Civil Engineering from the ESPOL Polytechnic University in Guayaquil. Since 1999 he has been working in consultancy firms in several areas such as foundation and pavement engineering, road drainage and road geometric design projects. In 2005, he joined a hydropower project in the Ecuadorian highlands as supervisor engineer. His main responsibilities were the supervision of cutting and filling earth volumes, the control on the construction of the water intake, weir, conduction pipes, reservoir, pressure pipes and power house as well as collateral works such as dikes, revetments, culverts and roads.

From October 2006 till April 2008 he pursued a MSc. program in Water Science & Engineering, specialization Hydroinformatics, at UNESCO-IHE, which was awarded with distinction. This post-graduate education was a scholarship from the Netherlands Fellowship Program (NFP). His thesis analyzed the quality improvement of remotely sensed imagery, by means of the information and appropriately selecting interpolation techniques to estimate missing data. The results enhanced an existing algae bloom model using in-situ data. The research was carried at the Deltares Institute (formerly Delft Hydraulics).

Following his MSc. studies, he undertook further research aimed to enhance his thesis' outcomes for publication, until October 2008. Since then, he became a PhD fellow participant within the WETWin project, 7th Framework Program FP (November 2008-October 2011). His main duties were enclosed in the work packages 3, 6, 7 and 8 dealing with data collection, organization, gap analysis, water systems modeling, expert elicitation and decision support system. He is member of the International Association of Hydraulic Research (IAHR) and the Ecuadorian Civil Engineering Association (CIEC).

## List of publications

Arias-Hidalgo, M., Bhattacharya, B., Mynett, A., van Griensven, A. (2012), "Experiences in using the TRMM data to complement rain gauge data in the Ecuadorian coastal foothills", Journal of Hydrology and Earth System Sciences, Discuss., 9, 12435-12461, doi:10.5194/hessd-9-12435-2012.

Arias-Hidalgo, M., Villa-Cox, G., van Griensven, A., Mynett A., (2012), "A simple pattern simulation in daily streamflow series", Journal of Hydrological Sciences (in revision)

Arias-Hidalgo, M., G. Villa-Cox, van Griensven, A., Solórzano, G., Villa-Cox, R., Mynett, A.E., Debels, P. (2012). "A decision framework for wetland management in a river basin context: the "Abras de Mantequilla" case study in the Guayas River

Basin, Ecuador", Journal of Environmental Science & Policy, Sp. Ed. (accepted for publication), doi:10.1016/j.envsci.2012.10.009.

Alvarez-Mieles G., van Griensven, A., Torres, A., Arias-Hidalgo, M., Mynett, A. (2012), "Relationships between aquatic biotic communities and water quality conditions in a tropical river wetland system", Journal of Environmental Science & Policy, Sp. Ed. (in revision).

Arias-Hidalgo, M., Villa-Cox, G., van Griensven, A., Debels, P., Mynett, A.E. (2011) "Decision Support system for wetland & catchment management based on modeling & expert judgment: Abras de Mantequilla wetland & Vinces-Chojampe subbasins in Ecuador", Stockholm Water Week, Stockholm, Sweden, August 2011.

Arias-Hidalgo M., Villa-Cox G., van Griensven A., Mynett A.E., (2011) "Integrating models in wetland - riverine systems", Joint Meeting of Society of Wetland Scientists (SWS), Prague, Czech Republic, July 2011, O-329.

Arias-Hidalgo, M., Villa-Cox, G., van Griensven, A., Mynett, A.E. (2011), "Filling missing data in streamflow series for supporting models". Valentine, M., Apelt, C., Ball, J., et al (Eds)., Proceedings of the 34th World Congress of the International Association for Hyro-Environment Research and Engineering : 33rd Hydrology and Water Resources Symposium and 10th Conference on Hydraulics in Water Engineering, Brisbane, Australia, 4016-4022.

Arias M., van Griensven, A., Debels P., Mynett A.E., (2010) "Comprenhending the wetland-river catchment integrated system". Proc. 8th International Symposium on Ecohydraulics, Seoul, Korea, Sept. 2010, 355-362.

Li, H., Arias, M., Blauw, A., Los, H., Mynett, A.E., Peters, S. (2010). "Enhancing generic ecological model for short-term prediction of Southern North Sea algal dynamics with remote sensing images." Journal of Ecological Modelling **221**(20): 2435-2446.

Li, H., Arias, M., Blauw, A., Peters, S., Mynett, A.E. (2009). A Pilot Study for an Enhanced Algal Spatial Pattern Prediction Using RS Images. Advances in Water Resources and Hydraulic Engineering. C. Zhang and H. Tang, (eds.), Springer Berlin Heidelberg: 738-743.

Arias, M., Li, H., Blauw, A., Peters, S., Mynett, A.E. (2009) "Enhancing Delft 3D-Bloom/GEM for Algae spatial pattern analysis: Filling missing data in RS images". Proceddings of the 8th International Conference on Hydroinformatics, Concepción, Chile, Jan. 2009, 1201-1210.

Li, H., Arias, M., Blauw, A., Peters, S., Mynett, A.E. (2009) "Enhancing Delft 3D-Bloom/GEM for Algae spatial pattern analysis: Model improvements". Proceedings of the 8th International Conference on Hydroinformatics, Concepción, Chile, Jan. 2009, 336-345.

T - #0423 - 101024 - C188 - 240/170/10 - PB - 9781138000254 - Gloss Lamination